WATER WAVES GENERATED BY UNDERWATER EXPLOSION

ADVANCED SERIES ON OCEAN ENGINEERING

Series Editor-in-Chief
Philip L- F Liu (*Cornell University*)

Advanced Series on Ocean Engineering – Volume 10

WATER WAVES GENERATED BY UNDERWATER EXPLOSION

BERNARD LE MÉHAUTÉ
SHEN WANG
University of Miami

World Scientific
Singapore • New Jersey • London • Hong Kong

Published by

World Scientific Publishing Co. Pte. Ltd.

P O Box 128, Farrer Road, Singapore 912805

USA office: Suite 1B, 1060 Main Street, River Edge, NJ 07661

UK office: 57 Shelton Street, Covent Garden, London WC2H 9HE

British Library Cataloguing-in-Publication Data
A catalogue record for this book is available from the British Library.

Library of Congress Cataloging-in-Publication Data
LeMéhauté, Bernard, 1927-
    Water waves generated by underwater explosion / Bernard LeMéhauté,
Shen Wang.
        p.   cm. -- (Advanced series on ocean engineering : v. 10)
        Includes bibliographical references.
        ISBN 9810220839 : $58.00. -- ISBN 9810221320 (pbk.) : $34.00
        1. Water waves.   2. Underwater explosions.   I. Wang, Shen.
    II. Title.   III. Series.
    TC172.L46   1995
    551.47'24--dc20   94-45506
                                         CIP

Printed in Singapore.

ACKNOWLEDGMENTS

This work was initially sponsored by the US Army Corps of Engineers, Coastal Engineering Research Center (CERC), Vicksburg, Mississippi, and funded by the Defense Nuclear Agency (DNA). The authors acknowledge the guidance and support of Lt. Camy Carlin, Lcdr. Harris O'Bryant, Lcdr. Kay Dinova, and Dr. Thomas Tsai of DNA, and Dr. Steven Hughes of CERC. Most of the experimental data was provided by the CERC under the direction of Douglas Outlaw. The authors are grateful for all the students who during their graduate studies have heavily contributed to the research presented here: Tarang Khangaonkar, Cha-Chi Lu, Sudhir Nadiga, Mills Soldate, Colby Swan, and Edward Ulmer. The authors also thank Dottie E. Mayol for her patience in typing the manuscript.

PREFACE

This book assembles in a single compendium the early 1990's state-of-the-art prediction of water waves generated by underwater explosions. Most of the literature on this rather specialized subject is scattered amongst a number of reports — some of them quite old — not always readily available. Also, there are times, as science progresses, where a synthesis is needed to select the best theories and results and eliminate unfruitful efforts. It is the ambition of the authors to achieve this goal and to establish a milestone upon which any future research can be based without risk of duplication or forgetfulness.

The content of this synthetical approach dwells upon years of research already described in a multitude of reports, but it mostly contains original contributions which the authors have carried out in recent years.

The subject of Explosion-Generated Water Waves (EGWWs) has not had as high a priority as compared to other explosion effects, e.g., shock waves. Therefore, progress has been slow and with few experiments due to the relatively limited effort dedicated to its resolve. Nevertheless, it is a subject which cannot be ignored considering the fact that high yield, near surface explosions are able to generate water waves of considerable magnitude. Our objective is to contribute to the wealth of knowledge on the effects and wave generation capabilities of underwater explosions so that future questions concerning the effects of EGWWs under various fictitious scenarios can be assessed.

The effects of the waves on various structures, fixed or floating, in deep water, on the continental shelves, on the coastlines or inside embayments and naval bases are also an important aspect of the subject. A realistic assessment of the relative importance of water waves is obtained by comparing these effects with other explosion-generated lethal effects. This work has also been done, at least partially. It was found that under some circumstances, the effect of water waves overcame the other effects. Indeed, water waves have a propensity for long duration and for traveling long distances. This allows them to affect structures in areas beyond the stand-off distance based on shock waves or other lethal effects.

The effects of EGWWs are very site-specific and will not be described herewith. This book is exclusively limited to the EGWW environment, more specifically, to the hydrodynamic phenomenology of wave generation and wave propagation. The effects of EGWWs in open water can be derived by direct application of the basic material presented herewith. Coastal effects, which include wave set-up, wave run-up and harbor response have been presented in a previous publication by one of the writers (Le Méhauté, 1971) and will not be reexamined here despite the increase in knowledge which has been achieved since.

In a sense, this book marks the end of an era, which the most fundamental problems of wave generation and propagation have been resolved with an accuracy allowing a realistic appraisal of their magnitude. The gaps that remain appear even more evident and a guideline is provided for the continuing research.

Our subject will be limited to the hydrodynamical phenomenology of wave generation and propagation. The physical behavior of these phenomena, the theories, and experimental data are described in detail. Predictive mathematical tools are developed allowing a quantitative assessment of the wave field as a function of various explosion parameters and the environment.

Chapter 1 is an overview of the wave generation process and the resulting wave field. The orders of magnitude of the physical characteristics of EGWWs are presented to assess their relative importance. Chapter 2 is entirely dedicated to the linear theory of impulsively generated water waves over a uniform depth. Chapter 3 deals with the calibration of the linear theory relating the hydrodynamics to the explosion parameters quantitatively. Chapter 4 describes the wave generation process in shallow water from a water crater reaching the seafloor and forming a cylindrical bore radiating outwards to the transformation of this cylindrical bore into a nonlinear wave. In Chapter 5, the sizing of the shallow water explosion-generated crater is determined as a function of yield from the analysis of experimental wave records. Chapter 6 discusses the dissipation processes taking place by wave-seafloor interaction. Chapter 7 deals with the propagation of EGWWs over nonuniform bathymetry. Chapter 8 details the production of EGWWs in the laboratory for experimental purposes. Chapter 9 presents the mathematical formulation of the Boundary Integral Element Method (BIEM) to investigate the dynamics of bubble formation which initiates EGWW formation. The results of a parametric analysis using BIEM are presented and comparisons with the empirical derivations are discussed.

The book is didactic in tone. Any reader with a fundamental background in hydrodynamics and water wave theories should be able to follow. The reader is led towards an understanding of the subject matter which should allow him or her to obtain quantitative predictions corresponding to various scenarios and/or pursue research and add further to our knowledge.

From a hydrodynamic point of view, the study of EGWWs extends beyond the classical wave theories and observed natural phenomena. Nevertheless, despite the uniqueness of this hypothetical man-made phenomenon and its fictitious implications, the study of EGWWs has proven to be most useful for understanding many natural phenomena taking place in the ocean. In this context, we hope that this book can be used beyond its apparent initial goal.

TABLE OF CONTENTS

Chapter 1

OVERVIEW OF THE WAVE GENERATION PROCESS AND THE WAVE FIELD

1. Explosion Parameters

A submerged explosion, like any other physical event that produces a localized disturbance of the water surface, generates a group of surface gravity waves that expand radially. The characteristics of Explosion Generated Water Waves (EGWWs) depend on a number of parameters characterizing the explosive and the medium.

The first parameter is the yield W. It is generally expressed in pounds of TNT equivalent for high explosives (HE). In the nuclear range the yields are expressed in kilotons (KT or 2 million pounds). The weight coefficient of HE to nuclear equivalent varies. The value of 0.8 is often used for practical purposes.

The second important parameter is the depth of burst, z, defined by the distance from the center of mass of the explosive to the free surface. The scaled depth of burst parameter is defined as $Z = z/W^{1/3}$, where W is assumed to be linearly related to a volume of TNT equivalent by its specific weight (dimension L^3). $Z = z/W^{1/4}$ and $Z = z/W^{0.3}$ have also been used.

The third parameter affecting the efficiency of the wave generation process is the water depth, d. If the water depth is sufficiently shallow such that the expanding gas bubble generated by the explosion interacts with the bottom, the wave field generation will be affected.

Based on the yield and the water depth, explosions are classified into three broad categories in terms of the depth parameter

$$D = \frac{d}{W^{1/3}} \quad (d \text{ is in feet and } W \text{ is in pounds of TNT equivalent}): \quad (1.1)$$

(1) Deep Water Explosion
$$D > 14,$$

(2) Intermediate Depth Explosion

$$1 < D < 14,$$

(3) Shallow Water Explosion
$$D < 1.$$

Qualitatively deep water explosions are those where the explosion crater on the water surface is small compared to the water depth. The waves created are analogous to those created by throwing a pebble in a pond and may, therefore, be described by linear theory at a certain distance from ground zero (GZ). In past literature (Le Méhauté, 1970 and 1971),

$$D > 6 \qquad\qquad (1.2)$$

was considered as the criterion for deep water explosion. Newer evidence (Wang *et al.*, 1991) infers that the wave generation mechanism is affected by larger values of the scaled water depth parameter than initially thought.

Shallow water explosions are those where the water depth is small compared to the explosion crater size. In this case, the ground is exposed and the wave generation mechanism is nonlinear, highly dissipative, and significantly different from the deep water explosions.

For very shallow water, the wave generation process is affected by ground cratering. The soil is left with a significant crater surrounded by a lip (Waterways Experiment Station, 1955). The corresponding debris projected into the atmosphere and falling randomly add considerable noise to the main wave field. Therefore, in theory, the wave generation process is also a function of the soil characteristics. At this time, this problem remains unresolved quantitatively and will not be considered here, even though some of the experimental data used for the calibration of mathematical models have been obtained on muddy bottoms. In the nuclear range, rocky sea floors would most likely behave like a muddy bottom from the wave generation point of view. In order to simplify the data analysis and to neglect the effect of ground cratering, the latest series of HE tests (Waterways Experiment Station, 1989) were performed on thick concrete slabs. It appears that the effect of soil response on the generation of water waves is the first gap which may require further research.

The last parameter of importance is the stand-off horizontal distance, r (in feet), between the location of the explosion or ground zero (GZ) and the point of observation. The stand-off distance is nondimensionalized with respect to water depth: $r^* = r/d$.

There are, of course, many other parameters that may need to be considered in the future, particularly those parameters concerning the explosion. The differences of energy density between HE and nuclear is quite large and the detonation process is, of course, completely different. It will be seen that, insofar as the water wave problem is concerned, both HE and nuclear explosions can be considered as point sources, and the general problem is simply reduced to an appropriate scaling of yield. In a word, given the same yield, the HE and the nuclear explosions will generate an identical wave field. This assumption has not been so much substantiated by experiments than by the application of theoretical and numerical hydrocodes. It implies that differences which may exist in the effect of the burst depth, z, for the two kinds of explosions are negligible.

Finally, the wave generation process may be affected by the bathymetry (bottom slope). Practically nothing is known about this effect. Wave propagation is also influenced by the bathymetry, a subject which will be covered in Chapter 7.

2. Overview of the Initial Conditions

When an underwater explosion occurs, the submerged detonation almost instantaneously produces a hot gas or plasma with a limited volume. High temperatures and pressures cause two disturbances of the ambient fluid:

(1) emission of a shock wave traveling outwardly,
(2) a radial motion of the fluid so that the "bubble" consisting of water vapor
     and explosive debris begins to expand.

Many of the fluid motions occurring in the case of small HE near-surface
explosions have been observed through reinforced glass windows and monitored
by high-speed cameras.   Their detailed complexity have been described by
Mader (1972), Craig (1974), and Kedrinskii (1978). Figures 1.1 and 1.2 provide
some insight on the form and evolution of various features. They demonstrate
the wide range of phenomena and variations of these phenomena as a function
of the depth of burst.

From this, it is clear that near-surface underwater explosions produce a
wealth of phenomena that are not all well understood. The limit of the "bubble" is ill-defined. A wide variety of gas clouds, upward moving jets, stems, and
"roots" at the bottom of the bubble occur. Therefore, it would seem that there
is little hope that such complex phenomenology can lead to well defined input
boundary conditions to determine the wave field. Whether or not these effects

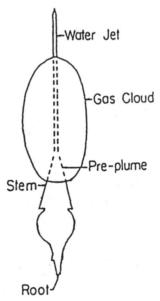

Fig. 1.1. Schematic of phenomena observed in near surface detonations (adopted from
Kedrinskii, 1978).

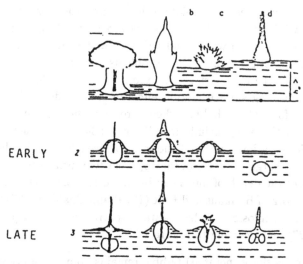

Fig. 1.2. Sketches of phenomena observed in near surface explosions (Kedrinskii, 1978).

would occur in the range of high (nuclear) yield remains an open question, but it is highly probable that they are not simple.

Extensive work was performed on bubble dynamics of both conventional and nuclear explosives (Cole, 1948; Snay, 1966; Holt, 1977). Later, small experiments were performed and families of hydrocodes were written and applied to underwater explosions adding a wealth of information to this earlier work. Recently, computer codes have been written based on the assumption that the fluid flow is irrotational and nondissipative, a subject which will be covered in Chapter 9.

In the case where the explosion SDOB is large (large Z), the bubble begins to rise due to its buoyancy. During the expansion phase, the pressure within the bubble falls considerably below the ambient hydrostatic pressure due to the outward momentum acquired by the water. The motion then reverses; the bubble contracts under hydrostatic overladen pressure acquiring inward momentum and adiabatically compressing the central gas volume to a second but lower pressure level. Upon reaching its minimum diameter, energy is radiated by the emission of a second shock wave, and the bubble expands again. The surface of the contracting bubble is extremely unstable — it may break irregularly forming a spray, and jets. The bubble may lose enough energy through repetitive expansions and contractions that it collapses entirely, leaving a mass of turbulent warm water and explosion debris, and therefore, no wave of any consequence will be generated. This case is typical for deeply submerged detonations ($Z > 8$) in deep water, and will not be discussed further since they do not present practical value for the problems addressed in this book.

In the case of the near-surface explosion, i.e., for smaller values of Z, the nature of the ensuing surface motion and the magnitude of the water waves depend on the form of the bubble when it reaches the surface. Depending on the value of Z, this may include a well-formed hollow column, a very high and narrow jet or a low turbulent mound, followed by the development of a prominent "base surge". Photos of the apparent free surface effects and plumes as a function of the depth of burst can be found in Cole (1948, plate XI).

The early work of Ballhaus and Holt (1974) and Chahine (1977) concerning near free-surface explosions in deep water is particularly descriptive. Quoting Chahine (1977) and referring to Figure 1.3:

> "This (explosion-generated) bubble remains approximately spherical during 70 per cent of its life, then collapses nonspherically and the upper side moves away from the free surface. However, this side reaches the opposite one before the formation of a microjet. During the growth of the bubble its upper side is attracted by the free surface and the bubble is elongated towards this surface, itself being disturbed regularly. Then the bubble collapses with the formation of a reentrant jet. A thin jet comparable to Taylor's instability arises from the free surface, while another microject penetrates the bubble, pierces it, and continues to be seen moving away from the free surface. This jet seems to carry away with it quantity of gases which later collapses, while two lateral jets are produced on the free surface on both sides of the first jet .... The jet formed on the free surface or "plume" appears a very long time after the explosion, during the collapse and not the growth of the bubble".

The explosive debris, mud and water are expelled upwards then fall randomly. The complex free-surface disturbance is at the origin of the wave generation process as described in the following.

If the explosion takes place in shallow water ($d/W^{1/3} < 1$) it is necessarily a near-surface explosion. The shallow water wave generation process is relatively simple and well understood. Initially, the water is expelled upwards and radially, forming a plume and a water crater with a watery rim or lip. The lip evolves into a small cylindrical bore which expands radially and decays rapidly to form a leading wave [Figure 1.4(a)]. The water crater exposes the soil to the atmosphere. After reaching its maximum size, the water crater collapses and the water rushes inwardly under the influence of gravity onto the soil crater, analogous to the dam break problem [Figure 1.4(b)]. The local reduction in sea level following the first wave crest caused by the inrush of water is transmitted outwards as a long shallow wave trough. When the water edge of the

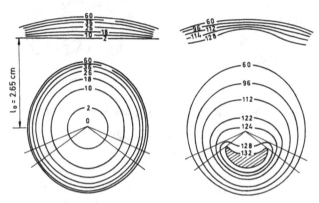

(a) Bubble No. 20: $R_{max}/l_0 = 0.60$, time scale: 13×10^{-5} s.

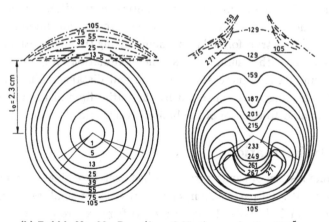

(b) Bubble No. 23: $R_{max}/l_0 = 0.95$, time scale: 8×10^{-5} s.

Fig. 1.3. Shape of the bubble and the free-surface at different times (Chahine, 1977).

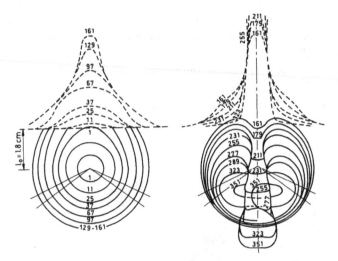

(c) Bubble No. 25: $R_{max}/l_0 = 1.5$, time scale: 8×10^{-5} s.

Fig. 1.3. (*Continued*)

inward motion reaches the center, a very high peak of water is thrown up in the center and a surge is formed which also expands radially on the top of the inward flow. This surge is analogous to a cylindrical bore, and dissipates a large amount (near 40%) of the potential and kinetic energy initially imparted by the explosion to the water [Figure 1.4(c)]. As the bore expands radially and decays [Figure 1.4(d)], it is transformed into a nonbreaking (nondissipative) undulated cylindrical bore, with a number of smaller undulations [Figure 1.4(e)]. The number of undulations increases with time and distance. The time interval between the leading wave generated by the initial lip and the following wave is directly related to the size of the limited crater and the initial water depth.

3. Cavity Formation

The time history of cavity formation has been investigated experimentally and theoretically. Of particular interest is a series of small scale experiments by Schmidt and Holsapple (1980) in a centrifuge, where gravity is a variable parameter. These experiments were performed under a variety of conditions and observed with a high-speed camera. The experiments included the cavity

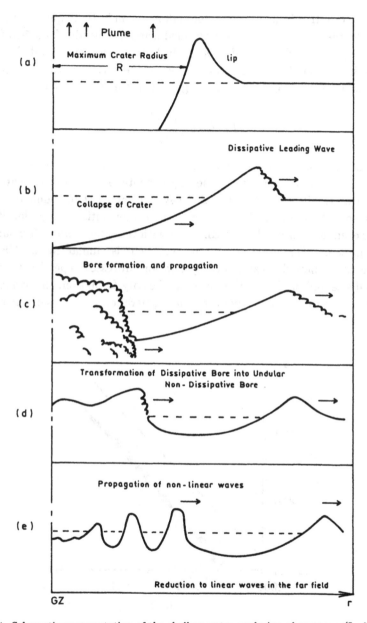

Fig. 1.4. Schematic representation of the shallow water explosion phenomena (Le Méhauté and Khangaonkar, 1991).

formation resulting from a drop of fluid falling on an initially quiescent body of fluid. The theoretical study was performed using hydrocodes for a 5-MT near-surface explosion. (Fogel *et al.*, 1983). In both experimental and theoretical studies, it was found that if T_f is the time for the surface cavity to expand to its maximum volume V_f, and g is the gravity acceleration, then

$$T_f \cong 0.8 \left(\frac{V_f}{g^3} \right)^{1/6}. \qquad (1.3)$$

A comparison of the calculated nuclear early-time $(t/T_f < 0.01)$ cavity growth for the 5-MT shot with the experimental HE results indicates that the nuclear results show faster scaled cavity growths than found in the HE experiments (Figure 1.5). This is due to the higher initial energy density in a nuclear event as compared to an HE event with the same yield. After sufficient cavity growth $(t/T_f > 0.01)$, the nuclear event is similar to the HE event (Figure 1.6). Since the water wave generation processes depend primarily on the final size and mildly on time history of cavity formation, this comparison gives credence to the similarity of the wave generation processes of nuclear and

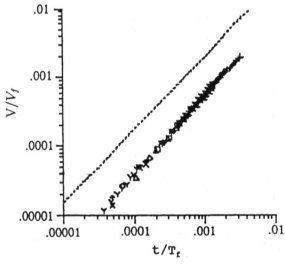

Fig. 1.5. Comparison of experiments (symbols) and calculation (dashed line) of early time cavity evolution (Fogel *et al.*, 1983).

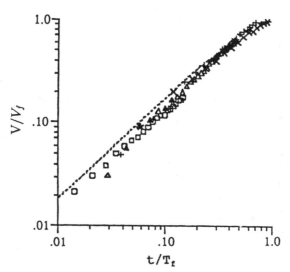

Fig. 1.6. Comparison of experiment (symbols) and calculation (dashed line) for late time cavity growth (Fogel *et al.*, 1983).

HE events, with appropriate scaling. Overall, the volume of the cavity as a function of time, Vol(t), is well approximated by the empirical relationship

$$\frac{\text{Vol}(t)}{V_f} = \sin \frac{\pi}{2} \frac{t}{T_f} . \tag{1.4}$$

The maximum volume V_f is an ill-defined, complex function of the explosion parameters, as will be seen in Chapter 9. In the 5-MT theoretical case, V_f is 1.15×10^8 m^3.

In shallow water, when the cavity depth reaches the seafloor, the radius of the cavity is well defined in a variety of sources. It is found that the horizontal radius of the cavity R_c is given by a simple relationship (Le Méhauté and Khangaonkar, 1991) developed in Chapter 5:

$$R_c = 4.4 \; W^{0.25} , \tag{1.5}$$

where R_c is in feet and W in pounds of TNT equivalent (Figure 1.7). The volume V_f is nearly equal to

$$V_f \cong 60 \cdot W^{1/2} d , \tag{1.6}$$

where d is the water depth in feet.

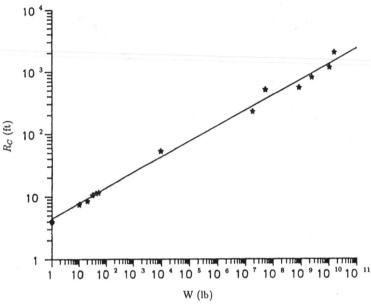

Fig. 1.7. The least squares fit through the numerical estimate of the physical crater radii by matching the recorded data with the crater collapse. The relation is given by Equation (1.5), $R_C \propto W^{0.25}$ (Le Méhauté and Khangaonkar, 1991).

4. Upper and Lower Critical Depths

Limiting our subject to the depths of burst that may produce a significant wave field, we will concentrate our effort to the cases where Z is smaller than about 10. Then the following is observed.

Experimental measurements indicate that the coupling between detonation near the free surface and the wave field vary significantly as a function of the depth of burst, depending on the water depth. For large depths of burst, the coupling variations can be explained by the phase of the bubble expansion when it reaches the surface. But for near-surface explosions in deep water, the variations cannot be explained by this effect.

Figure 1.8 illustrates this point. It is an attempt by Schmidt *et al.* (1986) based on the data of Pace *et al.* (1970) to group the various parameters relating to the maximum wave elevation (amplitude) $\eta_{\max}$ at a stand off distance r as a function of yield W, depth of burst z, and atmospheric pressure p. One

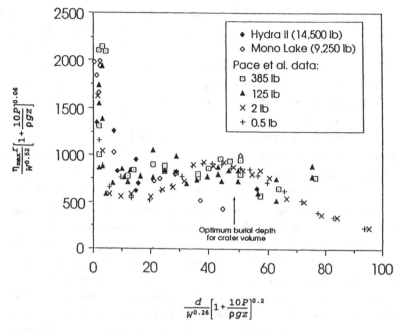

Fig. 1.8. Maximum wave height data from Pace *et al.* (1970). The data are plotted according to a scaling relation which presumes that atmospheric pressure is an important factor in determining wave height (Schmidt and Housen, 1987).

observes a mild peak when the scaled depth of burst Z is about 3. This has been called the lower critical depth (LCD). The LCD is clearly analogous to the influence of burst depth on crater dimensions in solid material. It is related to the balance between the explosion energy going into cratering and that vented into the atmosphere.

For very small values of depth of burst, the data exhibit considerable experimental scatter and peak in amplitude at a value varying between 0.5 to 2 times that at the LCD. This is called the upper critical depth or UCD.

The UCD is a rather puzzling aspect of wave generation. As pointed out by Hirt and Rivard (1983): "The experimental scatter suggests that the physical conditions for generating large waves at this depth involve a crucial balance of many phenomena. This should not be surprising considering the wide variety of interesting (and complicated) flow features that has been observed in tests .... Therefore it is not surprising that an understanding of the origin of the UCD and LCD is also lacking".

Another interesting feature which has been experimentally observed, but not clearly confirmed, is a change of phase of the leading wave radiating from GZ, depending upon whether the depth of burst is higher or lower than the UCD. This change, in fact, is predicted by theoretical models of wave trains generated by an initial impulse acting downward on the surface and by an initial surface elevation in the form of a crater, respectively. This suggests that the impulse model may be more appropriate for explosions above the UCD, and the water cratering model below the UCD.

There is also source evidence that the UCD may only exist for small yield (smaller than 300 pounds of TNT). The existence of the UCD is still somewhat in question for large explosions, since several attempts to reproduce it with 10,000 pounds. HE charges have been unsucccessful. It has been suggested (Kriebel, 1969) that the UCD effect is obtained from interferences between the direct incident shock wave and its reflected wave, resulting in more effective containments and greater cavity expansion than from deeper (or shallower) charges. As the detonation depth increases, the pressure impulse on the free surface has less and less effect on cavity formation and ultimately becomes negligible. This, according to Kriebel, would influence the shape of a theoretical cavity. In conclusion, the wave generation process at the UCD is a very sensitive function of the depth of burst, leading to various forms of cratering collapses. Additionally, the large data scatter obtained under fixed experimental conditions at the UCD are the result of a chaotic process, resulting from Taylor instability of the collapsing cavity.

The fact that less scatter occurs in the case of explosions in shallow water substantiates this hypothesis, because the collapse comes early from the side of the cavity and the sea floor forms a relatively fixed boundary. This seems to remain true as long as the sea floor is "clean dried" and exposed to the atmosphere during the early phase of wave generation. The decrease of coupling for explosion taking place at the surface is a result of the increase in energy vented into the atmosphere.

The basic question which remains, insofar as the EGWW is concerned, is whether or not the bubble dynamics, such as determined by hydrocodes, can be used as input boundary conditions for determining the wave field. In light of these physical observations, the problem is not an easy one to answer. The inherent limitations of the hydrocodes regarding the inclusion of turbulent dissipation processes alone present a severe limitation.

Due to the extreme selectivity of the experimental conditions which are necessary to reproduce the maximum coupling at the UCD, the maximum waves are difficult to reproduce and obtain operationally. They are only of interest for determining the maximum possible effect of EGWWs from a defensive point of view. The average values (AVG) obtained at the LCD should be used for a practical prediction for offensive effects of EGWWs, because these are not as sensitive to precise depths of burst.

5. The Compressible Hydrodynamic Phase

A theoretical mathematical model for the simulation of the hydrodynamics related to underwater explosion and subsequent bubble dynamics and free-surface effects was formulated. See, for example, Amsden (1973), Hirt and Rivard (1983), Fogel *et al.* (1983), and Mader (1988).

The corresponding "hydrocodes" are based on the Navier-Stokes equations for a compressible fluid, which can be treated numerically, given the appropriate equations of state and initial boundary conditions imposed by chemical or nuclear effects. These codes provide the most realistic, theoretical picture of an idealized free-surface time history and mass movement which is at the origin of the water waves generation process. The hydrocodes provide a wealth of information on the bubble dynamics. They have been very useful in establishing the hydrodynamic equivalence and scaling of nuclear and HE explosions. The results lend credence to the extrapolation of HE data to the nuclear range. But they also have their limitations as seen in the following section.

Whereas the duration of the initial shock wave generation is nearly instantaneous, the phenomena involving mass motion of the water and the duration of initial disturbances may be quite large, starting from the first appearance of a mound as the "bubble" nears the surface, to the collapse under gravity of the water thrown upwards in the form of a "plume". The gravity effects are not related to shock interaction and can be treated separately.

In the initial phase, the flow is compressible. As soon as the shock wave separates from the bubble front, the flow left behind behaves like an incompressible fluid subjected to the pressure of the gas or plasma inside the bubble. Compressible hydrocodes can be used as initial conditions to an incompressible fluid flow. This allows considerable simplification in the modeling of the water wave generation process.

6. The Incompressible Hydrodynamic Phase

Hydrocodes for incompressible fluids are just simpler versions of compressible codes. They allow for the investigation of the bubble dynamics and mass movement beyond the time duration of the compressible phase, where the shock wave separates from the surface of the bubble.

The hydrocode results compared quite satisfactorily with known experimental results. For example, Figure 1.9 was obtained by hydrocodes (Fogel *et al.*, 1983), and Figure 1.10 was obtained with a high-speed camera from the explosion of a very small yield against a glass window. Figure 1.11 compares

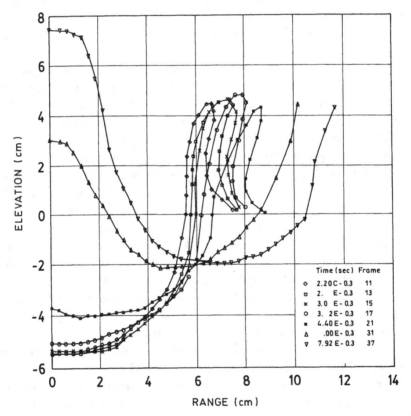

Fig. 1.9. Time histories obtained by hydrocodes of the free-surface profiles generated by a simulated near-surface explosion in deep water (Fogel *et al.*, 1983).

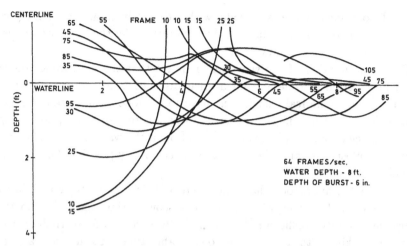

Fig. 1.10. Experimental cavity shape obtained by a small TNT explosion recorded by a high-speed camera (by courtesy of URS).

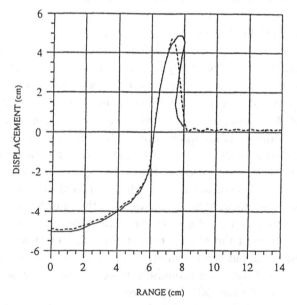

Fig. 1.11. Comparison of experimental trace (solid) and numerical representation (dashed) (Fogel *et al.*, 1983).

the results obtained by the hydrocodes with the experimental results at a given time. These figures clearly illustrate the validity of the theoretical approach. However, there are also limitations.

The determination of the wave field requires that the hydrocodes be run over a long period of time and for a great distance from the origin. At the same time, the numerical calculations have to be done with finite differences defined by a very small mesh. Because of this inherent difficulty, not enough cases have been done to allow parameterization of the wave field as a function of all the explosion parameters. This task is technically possible, but so far, it has not been done.

Additionally, hydrocodes have an inherent deficiency in their inability to assess the amount of energy dissipated by viscous turbulence. The choice of a suitable viscosity coefficient that realistically accounts for turbulent dissipation, by taking into account the considerable time and distance and the increase of temperature is beyond the state of the art. In practice, the value of the viscous coefficient is chosen for numerical stability, and it is a function of the mesh size. Therefore, any coincidence between the results of the viscous dissipation function given by hydrocodes and the energy dissipated by turbulence is purely fortuitous.

As previously indicated, the hydrocodes provide a fairly realistic picture of the energy partition going into the various effects, such as shock waves. But, since only a small amount of energy goes into the wave field, a relatively small error in the other effects (shock waves, etc.) would lead to large errors in the kinetic and potential energy at the origin of the water wave field.

For the EGWW problem, the objective is to obtain an input boundary condition for the definition of the water wave field, rather than to resolve the many associated problems of shock propagation, chemical or radioactive debris distribution, and thermal energy lost to heated or vaporized water. Theoretically, all the effects should be included in an effective model for analyzing the wave generation process, unless it can be shown that they can be neglected because they do not affect the wave characteristics. All these effects have indeed been investigated. Some of the studies neglect the effect of gravity, others make gross assumptions about the thermodynamic properties of the bubble dynamics and equations of state. Unfortunately, these effects cannot be assessed with sufficient accuracy for the EGWW problem.

This explains, at the outset, why the explosion bubble phenomenology as given by hydrocodes has been largely disregarded in the solution of problems concerned with surface-wave prediction methods, as will be seen in Chapters 2 to 5. This is remediable but requires much more research than has been dedicated to the problem.

A promising method, which is still in its infancy (as an application to the EGWW generation process) is the Boundary Integral Technique. This method is presented in Chapter 9 and will only be explained succinctly here in the context of its adaptability to describing the physical processes.

7. The Potential Flow Approach

If one assumes that the flow is incompressible, nondissipative, and irrotational, then it can be defined by a potential function. This approach has been used by a large number of researchers involved in bubble dynamic problems — due to explosion or otherwise (see, for example, Chahine, 1977).

The Boundary Integral Equation Method (BIEM) consists of determining numerically the solution to a potential flow problem. This method compares quite well with the incompressible hydrocodes in terms of accuracy. The method also provides a clear picture of the bubble dynamics as a function of initial conditions imposed by the time history of the pressure induced by the explosion (see Figure 1.12 from Wilkerson, 1988). The BIEM provides a frictionless solution whereas the hydrocodes provide a fictitious dissipative solution, so it is difficult to assess which method is the most accurate. The main advantage of the BIEM is its relative simplicity. The simplification introduced by the assumption of irrotationality leads to a computing time smaller by at least two orders of magnitude shorter than the hydrocodes. This permits to systematic treatment of all the parameters which cannot be done with the hydrocodes. Because the corresponding solution is nondissipative, one expects that the corresponding wave field will be overestimated. Nevertheless, it is reasonable to assume that the time history of the bubble boundaries provided by the potential flow solution yields adequate initial input boundary conditions for the determination of the wave field. The potential flow approach is the most direct and simplest and at the same time it provides as much accuracy as the incompressible flow hydrocodes. On the other hand, a short but important dissipative process which is not given by the BIEM must be introduced between the bubble dynamics and the wave field.

In the deep water case, dissipation can be introduced when the water wave exceeds a limited steepness near ground zero so that an upper limit is imposed

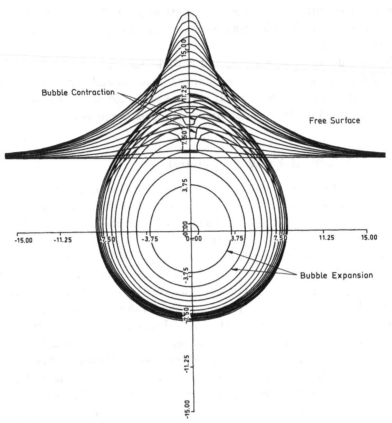

Fig. 1.12. Theoretical cavity shapes obtained by potential flow numerical calculations (Wilkerson, 1988).

to the wave amplitude. But this modification may not be sufficient because turbulent viscous dissipation processes also result from the crater lip projections and the fall of the mass of water contained in the plume. Therefore, a wave field based on a complete nondissipative potential flow approach provides results which run the risk of being slightly overestimated.

In the case of an explosion taking place in shallow water, satisfactory results are obtained deterministically. The nonlinear long wave approximation applies, and the powerful momentum theorem applied to a (cylindrical) bore allows a deterministic calculation of the energy loss. It will be seen in Chapter 4 that paradoxically, the shallow water case can be theoretically resolved practically in its entirety despite the complexities introduced by the nonlinear convective forces inherent to shallow water waves. In deep water, although the wave theory is simpler (i.e., the linear approximation is valid), it is not possible to resort entirely to a theoretical approach, unless the turbulent dissipative processes are negligible, which is not the case.

As is the case of the hydrocodes, this explains why the explosion bubble phenomenology has been largely disregarded in surface-wave prediction methods thus far. The difficulties encountered with incorporating turbulent dissipative processes resulting from the fall of the plume, projections, and gravity shock waves (bore and breakers) in the development of the codes are a major reason for this neglect.

8. The Prevailing Theoretical Approaches

Owing to the complexities of the previously described methods, and faced with the problem of providing a quantitative assessment of the magnitude of EGWWs, a mix of theory, experimental results, and empiricism is the favored approach. The problem consists of defining a treatable surface-wave generator as a function of the explosion parameters. Two approaches have been followed.

The first approach is based on the linear wave theory in which the dissipative and nonlinear convective terms are neglected. A generalized model is established which is valid for deep, intermediate, and shallow water depths. The model is calibrated by experimental results of wave records in the far field where the waves are nearly linear.

The second approach is nonlinear and dissipative. It is nearly fully theoretical, and is verified by near-field experimental results where the waves are nonlinear. This approach is valid in shallow water only.

In brief, the first approach assumes that the initial disturbance is defined by a (fictitious) linear localized perturbation in the form of either an impulse acting on the free surface $I(r)$, or an initial free surface elevation $\eta_0(r)$, or an initial vertical free-surface velocity $w_s(r)$ or any linear combination thereof. This is termed the "Black Box" approach in which all the complex phenomena taking place at GZ are ignored, and replaced by an idealized linear equivalent.

Because the wave motion is assumed linear and defined by a potential function $\phi(r, t)$, where r is the radial distance from GZ and t is the time, the initial conditions in terms of the potential function at $t = 0$, are

$$\phi(r)|_{t=0} = I(r_0) \tag{1.7}$$

or

$$\phi_t(r)|_{t=0} = \eta_0(r_0) \tag{1.8}$$

or

$$\phi_{tt}(r)|_{t=0} = w_s(r_0) \tag{1.9}$$

where the subscripts t and tt refer to the first and second differentiation with respect to time t.

Then, given an experimental wave record at distance r from GZ, the initial conditions are determined numerically by reverse transforms (Le Méhauté et al., 1989; Wang et al., 1989; and Khangaonkar and Le Méhauté, 1991).

For the sake of mathematical convenience, these initial conditions are matched by mathematical functions which are then quantitatively parameterized as a function of yield, depth of burst, and water depth. The best mathematical model obtained by this approach yields the initial disturbance as the sum of a parabolic water crater with lip, $\eta_0(r_0)$, and an upwards velocity, $w_s(r_0)$, in the form of a mound. Actually, the results obtained with the parabolic crater alone such as given by Figure 1.13,

$$\eta_0(r) = \eta_A \left[2\left(\frac{r}{R}\right)^2 - 1 \right] \qquad r \leq R,$$

$$\eta_0(r) = 0 \qquad\qquad\qquad r > R, \tag{1.10}$$

provide a fit to most of the data. Then only the two parameters, η_A and R, need to be defined as functions of the explosion parameters.

The limit of validity of the theory stems from the linear assumption, i.e., the Ursell parameter

$$U = \frac{\eta}{L}\left(\frac{L}{d}\right)^3 \ll 1, \tag{1.11}$$

where η is the maximum wave crest elevation and L is a typical wave length. This means that the theory is not a good description of the wave field near GZ or in shallow water. Of course, its description of the initial condition at GZ itself is purely fictitious, as implied at the beginning.

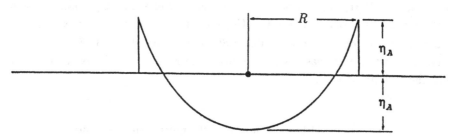

Fig. 1.13. Hypothetical static initial free-surface disturbance.

The second approach (nonlinear and dissipative) is valid in shallow water only. This method assumes that the motion at GZ follows the nonlinear long wave approximation, and it is dissipative due to the presence of a cylindrical bore. Initially, the disturbance is that of a crater reaching the sea floor with a lip [Figure 1.4(a)]. The size of this initial crater can be obtained by reverse calibration from wave records as done in Chapter 5. It can also be obtained theoretically by application of the BIEM to the bubble dynamics as proposed in Chapter 9.

Figure 1.4 describes a physical phase of the wave generation process. For shallow water explosions, the following hydrodynamic methods are used: (1) the collapse of the water is analyzed based on the nonlinear long wave theory, in analogy with the dam break problem, (2) the bore formation and propagation is investigated by application of the momentum theorem in analogy with the treatment of a dissipative tidal bore, and (3) when the bore height reaches a critical value, the initial disturbance evolves into a nondissipative, cylindrical, nonlinear wave defined by the Kordeweg and deVries (KdV) equation in a cylindrical coordinate system. The vertical acceleration is no longer neglected as in the long wave theory and the Ursell parameter is approximately unity. Due to the wide range of frequencies encountered in an EGWW train, the KdV equation has actually been extended to describe the higher frequency waves so that the theory remains valid from the long leading wave to the shorter trailing waves. The extended KdV (EKdV) equation is a recent (1990) original contribution developed specifically for EGWWs, which is also valid and applicable to all kinds of periodic or transient gravity waves whether plane waves or waves with cylindrical symmetry (Khangaonkar and Le Méhauté, 1991). By virtue of this extension, the EKdV equation overlays not only the range of conoidal wave (Ursell parameter: $U \approx 1$), but extends the validity into the range of the nonlinear Stokesian wave ($U \stackrel{\sim}{<} 1$). Eventually as time and distance increase, wave amplitude decreases, and the KdV equation matches the linear solution.

Table 1.1. Methodology. Linear wave theory for deep, intermediate and shallow water.

	Theory (nondissipative)	Experiments
GZ	Fictitous linear disturbance at time $t = 0$	Calibration measurements of free-surface elevation $\eta(t)$ at distance r.
Near field	Not valid.	
Far field	Linear wave solution.	Calibration measurements of free-surface elevation $\eta(t)$ at distance r.

Table 1.2. Methodology. Nonlinear wave theory for shallow water.

	Theory	Experiments
GZ Crater collapse bore formation.	Nonlinear long wave theory. Bore equation, dissipative.	Calibration measurements of free-surface elevation $\eta(t)$ at distance r.
Near field	Extended nonlinear KdV.	
Far field	Extended linear KdV.	Measurements of free-surface elevation $\eta(t)$ at distance r.

These two routes are schematically described in Table 1.1 for the linear theory and Table 1.2 for the nonlinear shallow water case.

9. Brief Historical Review of Past Investigations on the Linear Wave Field

The first theoretical treatment of wave generated by an initial free-surface disturbance of infinitely small radius but finite energy is due to Cauchy (1815) and Poisson (1816). The method is presented and extended by Lamb (1904, 1922, 1932) by use of Fourier's Integral Theorem in the one-dimensional case and Neuman's Theorem in two-dimensional cases. The solution of the movement of the free surface, in the form of an infinite series has limited physical significance because the input energy is implanted on an area of infinitely small extent. Kelvin's method of the stationary phase applied by Lamb, nevertheless serves to show that at large distances r from the original source or GZ, the wave amplitude of the maximum wave decays as $1/r$.

The generalization of Lamb's method to a case of an initial disturbance of finite extent was developed by Terazawa (1915) for cases of deep and intermediate water depth. Terazewa also investigated the effect of the depth of burst z in the case of an impulsive explosion. He found that the initial amplitude of the wave motion is proportional to $z^{-3/2}$. For a different model of explosion in which an abrupt pressure rise is followed by a gradual fall, Lamb (1922) deduced that the wave amplitude varies like z^{-2}.

Following this, much was learned from Japanese scientists as a result of their concern for tsunami waves, particularly Unoki and Nakano (1953a,b,c) who extended the work of Lamb and Turazawa where the initial disturbance is a uniform piston-like displacement of the surface over a circle of radius R. They found that the free-surface elevation exhibited beatings with modes determined by zero values of the Bessel function $J_1(kR)$. Unoki and Nakano also extended their work to the case of an initial impulse $I(R)$ of Gaussian form in which the solution exhibits an absence of beating and merely decays monotonically with time. A generalization of the Cauchy-Poisson Theory to a finite disturbance of an arbitrary form was developed by Le Méhauté (1963). Analytical solutions were also given for uniform circular and paraboloid free-surface disturbances in the form of a series.

Eckart (1948) demonstrated that the method of the stationary phase is not valid at the wave front in the one-dimensional case. By making use of higher order approximations, however, he showed that the leading wave may be represented by Airy's integral such that its wave length increases as the cube root of time and its amplitude decreases as the cube root of distance. By approximating the dispersion relationship to include only the long wave portion of the spectrum and limiting the source disturbance to a long narrow strip, Kajiura (1963) arrived at a similar result. This approximation was in fact equivalent to that of the Boussinesq Theory of dispersive long waves. The approximation of longitudinal dimensions of Kajiura led to the conclusion that the source distribution affects only the overall amplitude (but not significantly) on the broadside propagation at a long distance from the origin.

Reverting again to the more pertinent two-dimensional case, a solution for EGWWs was presented by Kranzer and Keller (1959) in which the influence of water depths was taken into account. The initial conditions were given either by an impulse or a free-surface deformation of finite size in the form of a paraboloid. Kranzer and Keller's derivations are asymptotic solutions based on the method of stationary phase, which are valid at a far distance from GZ. They follow transform techniques initially developed by Sneddon (1951) and Stoker (1957), which will be used in a following section. The approach of Kranzer and Keller has been retained to characterize EGWWs in deep and intermediate water depths ever since. Because the method is based on the stationary phase approximation, which is not valid for the leading wave, the theory of Kranzer and Keller is pertinent in cases where most energy is found in the trailing waves. This is the case of EGWWs in deep water at a long distance from GZ.

Much is to be learned from a paper by Kajiura (1963) on the leading wave. Although not developed for EGWWs, the Kajiura theory resolves the problem of the leading wave. In theory, his approach also allows waves due to time-dependent free-surface deformation to be determined. The theories of Kranzer and Keller and Kajiura have been used by a number of researchers of EGWWs (Whalin, 1965; Le Méhauté, 1970, 1971; Fogel *et al.*, 1983). So far, the main purpose of this research had been to determine the most suitable initial surface deformation used as input boundary conditions, which would yield a time history of free-surface elevation at a distance from GZ that resembled recorded experimental results. In the process, the sizing of the initial water crater in a parametric form as a function of yield, depth of burst, and water depth was determined. Also Le Méhauté (1970), Fogel *et al.* (1983) applied the theory of Kajiura to a time-dependent free-surface initial deformation. The Airy integral was integrated numerically by Whalin (1965) to determine the error resulting from the method of the stationary phase used by Kranzer and Keller. The method, however, is extremely complex numerically and cannot be generalized for practical use.

10. Main Physical Features of EGWWs as Seen Through the Linear Wave Theory

If the water is of constant depth, the wave field will show circular symmetry even though the initial disturbance is chaotic. The immediate analogy is that of a pebble dropped into a pond (Le Méhauté, 1971).

If the initial disturbance is thought of as a kind of "white noise" of all frequencies, many of the properties of the developing wave system may be anticipated. In particular, since the medium is dispersive for the propagation of gravity waves, it is expected that with time the waves will be sorted according to frequency, the longer waves running ahead and the shorter waves trailing behind. A curve of the transient wave period versus time at any location, therefore, will be monotonically decreasing.

The energy distribution among the generated frequencies will not be uniform. Intuitively, we may expect that the spectrum will peak near a frequency corresponding to a wave length of a small multiple of the initial disturbance radius. For large yield explosions, this radius is large so that the dominant wave can be expected to be rather long. On the other hand, the dimension of an explosion source (for realistic yields) is small compared to seismic sources responsible for tsunami generation. We then see that the study of EGWWs will cover a range roughly midway between wind waves (short) and tsunamis (very long). Therefore, considering the possible range of generated wave lengths, attention should be given to the effect of the bathymetry and wave soil interaction over a water depth which is considered deep water for wind waves. Wave energy displays trailing waves in the case of deep water, but tends to shift towards leading waves in the case of a large explosion taking place in shallow water.

In amplitude, explosion waves may exceed both tsunamis and wind waves which implies that nonlinear effects can be important as described in the following section. Table 1.3 gives the amplitude and period of the maximum wave as a function of yield and distances in the case of near-surface explosions in deep water. The "UCD" results yield the maximum possible coupling between the explosion and the water waves. The "average" (AVG) results correspond to the most probable in the case of a near-surface explosion ($Z < 6$).

The waves resulting from a free surface disturbance as defined by Equation (1.6) are shown in Figure 1.14, where the parameters $R = 1478$ feet and $\eta_A = 173$ feet. The case treated here corresponds to a 10-MT explosion in deep water. The wave pattern is given as a function of time t at three distances r from GZ. The theoretical wave pattern is generally well verified as shown in Figure 1.15 by the experimental wave record obtained from a 9,260-pound TNT explosion and the corresponding theoretical envelope. At a given location, the wave system appears as a series of wave trains, or modulated waves of monotonically decreasing wave period. All of the waves originate within the same central disturbance and are simultaneously released to propagate radially outward. If the water depth is uniform everywhere, or relatively very deep, the wave pattern will be perfectly circular and consist of concentric rings of crests and troughs, bounded by an intangible "front" that expands outward at the limiting velocity $c_0 = (gd)^{1/2}$ for free gravity waves. At any instant in time the radial separation between successive crests (wavelength) will be largest near the front and progressively smaller towards the center. All individual waves of the system will retain their identity, although the total number of waves present will increase with time as though they were being pulled like accordion bellows out of a black box that comprises the source region.

In general, no two waves of the system will be of the same size, nor will the amplitude of any wave remain constant with position and time. Within this ever changing pattern, the energy distribution among waves will be manifested by amplitude modulation of the wave train in a manner which is determined by the nature of the source, its distance from the point of observation, and the depth of water. As the pattern expands, the amplitudes of all the individual waves will diminish on average, because the wave system contains a finite and constant amount of energy which is diffused with increasing time or radial distance. This effect can be resolved into two factors: dispersion,

Table 1.3. Period (T sec) and wave height ($2\eta_{max}$ ft, m) of the maximum wave in a wave train generated by explosion at upper critical depth (UCD) and most probable (AVG) in deep water at various distances.

Yield W TNT Equivalent (1 ton = 2,000 lbs)	T sec		$2\eta_{max}$ ft / m									
			5 nm		10 nm		50 nm		100 nm		500 nm	
	UCD	AVG	UCD	AVG	UCD	AVG	UCD	AVG	UCD	AVG	UCD	AVG
1 KT	14.7	15.6	2.98 ft.	1.32	1.5	0.66	0.3	0.13	0.41	0.06	0.03	0.01
			0.91 m	0.40	0.46	0.20	0.09	0.04	0.12	0.02	0.011	0.00
10 KT	20.8	22.0	10.4	4.62	5.2	2.32	1.04	0.46	0.5	0.22	0.1	0.05
			3.17	1.41	1.58	0.71	0.32	0.14	0.15	0.07	0.03	0.01
100 KT	29.3	31.0	36.0	16.00	18.0	8.00	3.6	1.60	1.8	0.80	0.36	0.16
			11.0	4.87	5.5	2.44	1.1	0.49	0.55	0.24	0.11	0.5
1 MT	41.5	44.0	125.1	55.5	62.4	27.7	12.58	5.5	6.2	2.8	0.62	0.55
			38.1	16.9	19.0	8.44	3.83	1.7	1.9	0.85	0.19	0.17
10 MT	58.6	62.1	432	192	216	96.0	42.1	24.4	21.6	9.6	4.3	2.4
			132	58.5	65.8	29.2	12.8	7.43	6.6	2.9	1.3	0.7
100 MT	82.7	87.6	1500	666	750	333	150	66.6	74	33.3	15.0	6.6
			457	202	228	101	46	20	22	10.1	4.6	2.0

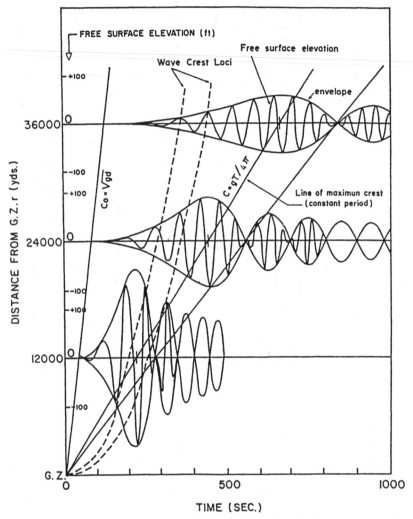

Fig. 1.14. Schematic drawing of water wave trains as function of time at three different locations (12, 24, and 36 thousand yards from GZ). The yield W is 10 MT, the depth of burst = 100 feet, the water depth d = 2000 fathoms (Le Méhauté, 1971).

due to the increase in wavelength and the number of individual waves; and geometric expansion, caused by the increase of crest length necessary to circumscribe progressively larger radii.

Most of these features are well illustrated in Figure 1.14 which shows three successive stages in the development of an explosion-generated wave train. The three theoretical computer-generated oscillatory curves of Figure 1.14 show the amplitude-time histories of a wave train generated at the origin as they would be recorded at the distances of 12, 24, and 36 thousand yards, respectively. The symmetrical curves bounding the wave trains comprise the wave envelope, and serve to define the distribution of energy within the train. The precise shape of the envelope depends upon the initial source conditions, whereas the space-time coordinates of the individual waves are independent of the source and depend only on the water depth. A characteristic of the wave envelope is that any identifiable portion of it, say, a node or antinode, propagates at uniform group velocity, as shown by the straight lines *o-a* connecting the origin with the nodal points delimiting the beginning (wave front), and *o-b* ending points of the first envelope maximum in Figure 1.14. The space-time trajectories of all waves of the system are curves, concave upwards because the waves are continuously accelerating towards the limiting phase velocity $c_0 = (gd)^{1/2}$ of the wave front. Thus, the waves travel faster and pass through the successive nodes of the wave envelopes, and therefore there are progressively more waves in each envelope segment with increasing time or distance. For a large explosion in the deep ocean, by the time the wave system has traveled a distance equivalent to 300

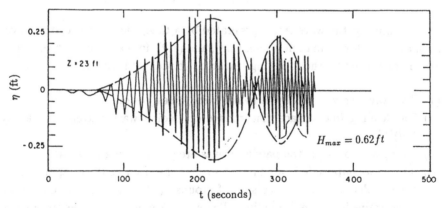

Fig. 1.15. Typical correlation between experimental results and mathematical model. Yield, 9620 pounds TNT; depth at exploson, 125 feet; depth at gage, 96 feet; distance between gage and SZ, 3600 feet (Le Méhauté, 1971).

water depths, there would be more than 100 waves between the front and the first nodal point.

A second important feature of the wave envelope is that its amplitude, as measured along any straight line through the origin of r-t diagram presented in Figure 1.14, is inversely proportional to its distance from the origin. Thus, the height of the highest wave in the upper wave train (Figure 1.14) is about one-third that of the corresponding wave in the lower train.

In theory the above features are related to the linear dimensions η_A and R of the original crater. In particular the maximum amplitude of the wave envelop η_{max} is given by

$$\eta_A R = 1.63\eta_{max}r \, . \tag{1.12}$$

The period of the maximum wave T is also of particular interest. It remains constant as a function of the distance from GZ and is given by

$$T = 2\pi \left[\frac{R}{4.2g}\right]^{1/2} . \tag{1.13}$$

The time duration between the wave front which travels at a speed of $(gd)^{1/2}$ and the wave of maximum amplitude which travels at a speed of $gT/4\pi$ is

$$\Delta t = r \left[\frac{1}{(gd)^{1/2}} - \frac{4\pi}{gT}\right] . \tag{1.14}$$

In summary, the wave trains generated by a theoretical linear disturbance, such as defined by Equations (1.7), (1.8), and (1.9), for example, in deep water are characterized by the following features (Le Méhauté, 1971):

(1) The waves travel radially outward from the explosion.
(2) The leading free surface disturbance or leading wave travels at velocity $(gd)^{1/2}$.
(3) At a given location, the passing waves are of increasing frequency.
(4) The wave amplitude is modulated so that the train of waves appear with time or distance as a succession of groups. (The wave envelope of those wave trains is a Bessel function. The waves are defined by the product of a Bessel function and cosine.)

(5) The number of waves in a given group increases with time or distance traveled.

(6) The length of a group increases with time or distance traveled.

(7) The frequency associated with a specific crest decreases with time or distance traveled (equivalently, a given crest moves forward within a group).

(8) The frequency associated with the maximum amplitude of a given group is constant.

(9) The maximum height of a given group decreases with time or distance traveled.

(10) The maximum height of successive groups passing a given point decreases.

(11) The maximum wave height varies inversely with distance from the explosion.

(12) The product of the maximum wave height and distance is simply related to yield and depth of burst.

All these features have been relatively well verified experimentally. Figure 1.15 is a typical comparison between an experimental wave record with the corresponding theoretical wave envelope.

In shallow water, the features presented in the deep water case remain generally valid, except for the fact that the leading wave takes on more importance (Figure 1.16). There are fewer waves per wave group, so the modulation in amplitude shown in Figure 1.16 is much less apparent. Close to GZ, there may be only one or two waves in the first wave envelope, so that the maximum of

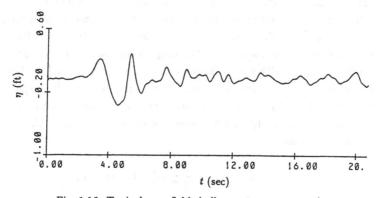

Fig. 1.16. Typical near-field shallow water wave record.

the envelope is ill-defined. The most significant feature is that the dispersion law is different in deep and shallow water. In shallow water, the leading wave, which tends to be the largest, decreases in height with the distance as $r^{-2/3}$ instead of r^{-1} as in the deep water case. More importantly, nonlinear effects start to show importance, so that the linear wave model becomes less valid, especially in the near field. The sinusoidal appearance of the wave profile is replaced by a conoidal profile, which exhibits high peaky wave crests separated by long wave troughs.

Despite these limitations, the linear methodology used for predicting waves generated by underwater explosions in deep water is remarkably simple. What is really surprising is that such a simple model works as well as it does considering the great complexity of the process. In particular, the linear wave theory based on the static water crater model is perfectly suitable, even though improvements can be obtained by using more complex initial disturbances, as will be seen in the shallow water case presented in Chapter 4.

11. Data Sources and Scaling

The experimental data used to verify and calibrate the mathematical predictive EGWW model, varies considerably in range of yield and in quality.

The largest bulk of data is found in a series of reports on experiments done at the Waterways Experiment Station (WES), Vicksburg, Mississippi, in the 1950s and 1960s. Some data of nuclear tests at Bikini Islands are also available. Most recently, a series of shallow water and intermediate water explosion tests were conducted at WES under the sponsorship of the Defense Nuclear Agency. A summary of the data available for empirical correlation is listed in Table 1.4.

The correlation was intended to cover both shallow and deep water explosions. As indicated earlier, one significant difference between shallow water and deep water wave propagation is the dispersion law. In deep water, wave height varies inversely with radial distance r, as a combined result of frequency and radial dispersion. In extremely shallow water, however, the large leading wave is expected to behave like a nondispersive solitary wave as a limit and its height should vary inversely as $r^{2/3}$ instead of r. It follows that the following decay law should hold

$$\eta r^\alpha = \text{constant} \qquad \frac{2}{3} \le \alpha(d, W, z) \le 1, \qquad (1.15)$$

where α is a function of the water depth d, yield W, and submergence z. The decay ratio α is to vary from 2/3 to 1 for shallow to deep water, respectively, and η is the maximum elevation of the leading wave.

Table 1.4. Data source.

Experiment	Date	Yield	Water depth (ft)
WES	1955	0.5–2048 lbs	0.5
WES	1960–1965	0.5–385 lbs	20–100
Mono Lake	1965	9250 lbs	10–100
WES	1986	10–50 lbs	1–2.7
WES	1988	10 lbs	1–4
WES	1989	0.5–10 lbs	1–5

In order to verify this relationship, Wang *et al.* (1991) and Khangaonkar and Le Méhauté (1991), developed an inverse method by which the initial conditions defined by the sum of a free-surface deformation and initial free-surface velocity, are determined numerically from measured wave records.

In the most simplified static form of the initial disturbance, the size of the water crater is defined by its radius R and its depth η_A (1.10). These are related to yield W, depth of burst z, and water depth d. If λ is a convenient geometrical linear scale between two explosions, the scaling problem is as follows (Johnson, 1959): The energy released by a conventional explosive (HE) is proportional to its volume, i.e., λ^3. When the charge is detonated, the gases expand and create a cylindrical crater; much of the water originally in the crater is pushed above the original water level in a plume and around the rim forming a base-surge which is the origin of the first outward traveling circular wave front. If L is designated as a characteristic length of a given explosion (i.e., $\lambda = L_1/L_2$ for two explosions 1 and 2), the scaling law may be expressed by the following dimensionless parameter:

$$\pi_1 = \frac{pL^3}{W}, \tag{1.16}$$

where p denotes the overburden pressure exerted on the crater boundary at the cavity-fluid interface and W is yield in terms of energy.

On the other hand, the total energy of the surface wave as required for geometrical similarity must be proportional to $\eta_A^2 R^2$, the potential energy of the crater, and must follow a quadruple scaling λ^4, which may be expressed by another dimensionless parameter

$$\pi_2 = \frac{\rho g L^4}{W}, \tag{1.17}$$

in which ρ is the fluid density.

There is experimental evidence (see Chapter 3) that, in deep water, η_A is proportional to $W^{1/4}$, whereas the radius R is proportional to $W^{1/3}$. This means that similitude is not possible: The EGWW generated by a large yield cannot be simulated at a small scale by a small yield. The small linear crater equivalent due to the small yield is a distorted (vertically exaggerated) model of the crater generated by a large yield.

If one reproduces the magnitude of the wave envelope in similitude, which requires that:

$$\lambda = \frac{\eta_1}{\eta_2} = \left(\frac{W_1}{W_2}\right)^{1/4}, \tag{1.18}$$

then the wave length, and time histories which relate to R are not in similitude.

The second option is to simulate the radius

$$\lambda = \frac{R_1}{R_2} = \left(\frac{W_1}{W_2}\right)^{1/3}, \tag{1.19}$$

but now the amplitude of the wave envelope is not reproduced.

In the case of shallow water explosions when the water crater reaches the sea floor, it is found that the radius of the crater R_c is proportional to $W^{1/4}$. Then similitude is possible as

$$\lambda = \frac{d_1}{d_2} = \frac{R_{c1}}{R_{c2}} = \left(\frac{W_1}{W_2}\right)^{1/4}. \tag{1.20}$$

This explains the complexity of the calibration and the necessity for a large number of tests to obtain a realistic quantitative prediction of EGWWs as explained in Chapters 3 and 5.

12. Propagation of EGWWs Over Nonuniform Water Depths

The slope of the sea floor is generally gentle enough that the wave generation process is a function of water depth at GZ, but is practically not affected by bottom slope. Conversely, the wave field is deeply affected by the bathymetry.

The leading wave traveling at speed $(gd)^{1/2}$ is affected by the bathymetry even in the deepest ocean. The following waves (and in general, the wave of maximum amplitude) "feel" the seafloor when the water depth d becomes smaller than half the deep water wave length, i.e., when

$$d \stackrel{\sim}{<} \frac{g}{\sigma^2} = \frac{gT^2}{4\pi}, \tag{1.21}$$

where σ is the transient frequency and T the transient wave period.

Referring to the wave periods presented in Table 1.3, for many hypothetical cases of near-surface explosions in offshore deep water, even the trailing waves are affected by the sea floor as soon as the waves reach the continental slope. For example, the period of the transient maximum wave due to a 1-KT explosion in deep water is 15.6 seconds, corresponding to a transient wave length of 1,246 feet which begins to be affected by the bottom as soon as the water depth is less than about 600 feet.

For shallow water explosions on the continental shelf, in embayments or harbor basins, the waves are affected by the bathymetry as soon as they radiate from GZ. Therefore, the wave field no longer follows radial symmetry. The wavefield is modified by refraction and energy dissipation due to wave-seafloor interaction and ultimately by wave breaking.

When EGWWs originating in the deep ocean approach the continental slope, a small part of the energy of the leading waves is reflected seaward. A fairly accurate estimate of the reflection coefficient can be made by assuming that each wave behaves as a periodic wave of the same period. From a hydrodynamic viewpoint, continental slopes are so gentle and wavelengths so short that, for most practical purposes, the methodology based on the conservation of energy flux is valid and reflections can be ignored (Le Méhauté, 1971). (This is not the case for the longer tsunami wave.) Under the assumption of the linear wave theory, the energy flux per unit length of wave crest is

$$F = \frac{1}{4}\rho g \eta^{*^2} C_g, \tag{1.22}$$

where ρ is water density, g is gravity acceleration, η^* is the amplitude of the wave envelope, and C_g is the group velocity. Accordingly, for a wave arriving perpendicularly to bottom contours, the wave amplitude η^* is modified by a shoaling coefficient K_S such as

$$\eta^* = \eta_0^* K_S \, , \qquad K_S = \left(\frac{C_{g0}}{C_g} \right)^{1/2} , \qquad (1.23)$$

where the subscript 0 refers to the deep water value.

At the edge of the outer continental shelf where $d \cong 300$ feet, K_S is given by a linear shoaling wave theory as a function of the wave period (Table 1.5) (Le Méhauté, 1976). In the range of explosion waves described in Table 1.3 with periods ranging from 10 seconds to 70 seconds or more, the increase of wave height due to shoaling remains small. If the wave height $H = 2\eta^*$ is large enough to break on the outer continental shelf or on the continental slope (i.e., for $H/d \stackrel{\sim}{>} 0.78$ or $H \stackrel{\sim}{>} 230$ feet), a nonlinear wave shoaling coefficient must be used, giving a larger wave shoaling effect as shown in Figure 1.17. The shoaling coefficient K_S is then determined by application of the principle of conservation of energy flux by taking into account nonlinear convective effects (Le Méhauté and Wang, 1980). Note that in Figure 1.17, L_0 is the deep water linear wave length, $L_0 = gT^2/2\pi$, H_0 is the equivalent deep water wave height, obtained by reverse calculation, and H_0/L_0 the deep water wave steepness (Svendsen and Brink-Kjaer, 1972). A more accurate breaking index than the one given by the limiting wave is given in the following as a function of bottom slope.

Table 1.5.   Linear shoaling coefficient K_S for 300 feet (100 meters) water depth vs. wave period (note $H = 2\eta^*$).

T (sec)	$K_S = \dfrac{H}{H_0}$
10	1
20	0.91
30	0.98
40	1.084
50	1.19
60	1.288
70	1.375

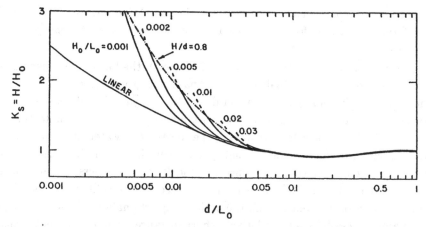

Fig. 1.17. Nonlinear shoaling of EGWWs (Svendsen and Brink-Kjaer, 1972).

In reality, as the EGWWs originating in deep water approach the continental slope and pass into shallow water, they continue to disperse both radially and angularly, even though the dispersive effect due to frequency is now reduced by the water depth. In addition to wave shoaling, the waves also refract and tend to become parallel to the bottom contours.

The propagation of nonperiodic waves over a varying bathymetry is based on simple theoretical principles of conservation of energy flux or conservation of wave action (see Chapter 7). The wave rays are defined by application of Snell's Law for wave refraction. Since the wave periods vary continuously with time and space, the wave rays change continuously at any given location. The wave energy flux also changes continuously because both the group velocity and wave amplitude change with time and distance. Because of the time dependency, and the nonlinearity of shoaling and dissipative processes due to wave seafloor interaction, the mathematical implementation of these relatively simple principles could be extremely complex and has been partly developed only recently. Approximate methods have been developed offering great simplification without significantly impairing the accuracy of results.

As a dispersive wave train (i.e., a train which consists of a series of waves of different lengths or periods, and thus different group velocities) arrives in shallower water, the individual wave amplitudes tend to become larger and their length shorter as the energy increment within each wave is concentrated in an increasingly smaller volume of water. This effect opposes the tendency for waves to become smaller due to dispersion and geometric spreading, and therefore, there exists for each wave a minimum amplitude at some point in its history.

In very shallow water $(d/L < 0.05, L \cong T \sqrt{gd})$, the linear wave dispersive effect tends to become negligible. Linear phase and group velocity approach a common value depending on the water depth only: $\sqrt{gd}$. The influence of wave period disappears and each wave keeps its identity. An approximate theory (used until 1983), consists of matching the linear deep water dispersive wave solution with the shallow water nondispersive solution, assuming that all the waves arriving at the edge of the continental shelf are nondispersive and behave like quasi-periodic conoidal waves or like a succession of solitary waves. This simplified approach, even though theoretically questionable, has proven to be reasonably successful over a relatively short distance. The subsequent transformation of the wave on the shelf is then treated wave by wave (Houston and Chou, 1983). The method described in Chapter 7 overcomes these limitations.

Eventually as water depth decreases and wave growth continues, the local wave amplitude will amount to an appreciable fraction of the water depth. In this "shallow water" regime, additional modification of the wave system is caused by amplitude dispersion. This wave amplification is nonlinear as convective acceleration effects become more important. Subsequently, the phase speed in very shallow water is also a function of wave amplitude, i.e., a higher wave tends to travel faster than a smaller wave of the same length. More importantly the wave height increases faster than predicted by the linear wave theory.

In shallow water, the wavelength as well as the phase speed also becomes a function of the wave height, and therefore, not only wave shoaling but also wave refraction exhibits nonlinearity. The wave then becomes "amplitude dispersive" in addition to being "frequency dispersive". However, nonlinear refraction effects generally have been ignored so far, owing to their complexity and their relatively small overall effect on wave amplitude. Also due to convective inertia, the individual wave profile may become unstable, each wave subdividing into a succession of two or three undulations which travel either as solitons or as undular wave packets separated by long flat troughs (Benjamin and Feir, 1967).

In the case where the explosion takes place in shallow water, the basic principle of wave propagation remains the same as previously described, i.e., the waves are also affected by wave refraction. However, the difficulty is compounded by the fact that this effect occurs near GZ and the waves are also affected by the nonlinear convective forces.

So far the problem is resolved in the linear case, at some distance from GZ using an asymptotic expansion. The method will be described in Chapter 7. The nonlinear nearfield waves in shallow water has been resolved for a uniform water depth only.

13. Dissipative Processes by Wave-Seafloor Interaction

As soon as waves generated in deep water reach a continental slope and interact with the seafloor, various energy dissipation processes occur. The dissipative processes due to wave-seafloor interaction are small on a steep continental slope such as encountered on the West Coast of the US. But they cannot be neglected on long shelves such as on the East Coast, or in the Gulf of Mexico, for example. They are of fundamental importance on the exceptionally long shelf of the Bering Sea.

For gentle slopes, the increase of wave height by shoaling is overcome by the decrease of wave height by energy dissipation (in addition to the remaining dispersive effects). The reasons why this effect is relatively important are twofold:

(1) Explosion generated waves are relatively long (as compared to wind waves) and, therefore, interact with the seafloor in much deeper water than wind waves.
(2) Most energy dissipation processes are nonlinear. The larger the wave amplitude, the more important the wave damping is. EGWWs are of much larger amplitude than tsunami and storm waves.

These dissipative effects have been formulated and presented in Chapter 6. The present developments are based on a combination of theories, laboratory data, and field measurements, mostly in the range of wind waves. Therefore, their application to the relatively larger range of EGWWs is obtained by theoretical extrapolation. These phenomena are not scalable, and in this respect, there are no existing phenomena in nature which parallel EGWWs. A maximum bottom velocity of 30 feet/second at the limit of a 6-feet boundary layer is conceivable, inducing a shear on the seafloor which remains unheard of under any observed natural phenomena.

Typical values of the maximum particle velocity under the wave crest of an EGWW at the still water level (SWL) and near the seafloor are presented in Table 1.6. These results have been obtained by the application of a nonlinear periodic wave theory over a horizontal seafloor (Colby, 1984). Particle acceleration remains, of course, smaller than the gravity acceleration g as in any gravity wave. The pressure fluctuations are also very large since they are of the same order of magnitude as $\rho g H$ (ρ is the density, H is the wave height). Furthermore, the results are a function not only of the bathymetry, which is generally well-defined, but also of the soil and subsoil physical characteristics of the bottom. This information is site specific and is not well-known in general.

Table 1.6. Particle velocities near the seafloor and at still-water level under the wave crest (maximum value) as a function of water depth d, wave period T, and wave height H obtained by the application of the parameterized nonlinear wave theory (Le Méhauté *et al.*, 1984).

Water depth $d\,^{ft}_{(m)}$	Wave period T sec	Wave height $H\,^{ft}_{(m)}$	U bottom ft/sec (m/sec)	U still-water level ft/sec (m/sec)
300 (91.4)	30	200 (60.9)	26.7 (8.2)	41.2 (12.5)
		100 (30.5)	15.6 (4.7)	22.2 (6.8)
	70	200 (60.9)	33.4 (10.2)	43.5 (13.3)
		100 (30.5)	21.5 (6.6)	26.1 (7.9)
200 (60.9)	30	100 (30.5)	20.3 (6.2)	28.4 (8.7)
	70	100 (30.5)	24.6 (7.5)	30.6 (9.3)
100 (30.5)	30	50 (15.2)	16.1 (4.9)	21.1 (6.4)
	70	50	20.6 (6.3)	25.8 (7.8)

The wave motion and wave induced pressure fluctuations on the seafloor cause three dissipative processes to occur (Colby, 1984):

(1) A turbulent boundary layer is developed over a movable bed (Carstens *et al.*, 1969; Vitale, 1979; Grant and Madsen, 1982).
(2) Coulomb friction in the soil sublayers is generated as a result of induced wave pressure fluctuations in the soil (Yamamoto *et al.*, 1978; Yamamoto, 1983; Takahashi, 1982).
(3) Energy is lost as a result of pressure-induced fluid flow in the permeable bottom soils (Putnam, 1949; Reid and Kajiura, 1957).

The relative importance of these three effects depends on bottom soil characteristics. (1) is generally the most important on a sandy shelf, but (2) is more important on a clay bed (silt or ooze, as encountered at the mouth of the Mississippi River, the Amazon, the Yangtze), and (3) is more important on coarse sand.

The energy loss in a turbulent boundary layer is a function of the relative roughness of the seafloor. In the case of a movable (sandy) seafloor, in addition to the loss of energy resulting from the relative roughness, the bed load regime also influences the energy dissipation.

One distinguishes three movable bed regimes: (1) the equilibrium sand ripple regime with alternate vortices on each side of ripple-crests, (2) the post vortex regime as the ripple wavelength increases and sand ripples are flattened out, and (3) the sheet flow regime, for which the flow entrains and transports layers of sediment, contributing heavily to energy dissipation.

For EGWWs, all three cases can occur. For large amplitude waves, the sheet flow regime prevails and a considerable amount of sediment is moved. The bottom velocity can easily reach a very large value, say 30 feet/second (10 meters/second) at the limit of the boundary layer which may be 6 feet (2 meters) thick. Accordingly, the wave-induced shear on the seafloor can be extremely large.

The results which are relatively well determined for the equilibrium sand ripple regime and post vortex regime but rather ill-defined in the sheet flow regime, which, as mentioned above, is the most probable regime in the case of large EGWWs. It is the least well-defined because EGWWs are not encountered in nature. Damping coefficients due to boundary layer effects have been determined using the best available sources of information in the oceanographic and hydraulic literature with theoretical extrapolation.

Table 1.7. Wave damping coefficient by turbulent boundary layer over a moveable bed. $[D_B \ (\text{m}^{-1}) \times 10^6 \ (SG = 2.6).]$

Water depth $d^m_{(ft)}$	Wave height $2\eta^m_{(ft)}$	Fine sand $D = 0.1$ mm Wave period (sec)		Medium sand $D = 0.5$ mm Wave period (sec)	
		40	70	40	70
30 (98)	10 (32)	30	24	20	30
50 (164)	10 (32)	9.4	7.7	5.3	4.3
	40 (131)	66	52	64	50
70 (229)	10 (32)	4.3	3.6	2	1.6
	40 (131)	30	24	28	23
90 (295)	10 (32)	2.4	2.1	.79	70
	40 (131)	17	14	15	13
	70 (229)	37	30	37	30
110 (360)	10 (32)	1.5	1.3	1.6	0.24
	40 (131)	10	8.9	9.2	4.9
	70 (229)	23	19	22.0	19

As previously indicated, the boundary layer damping coefficient, D_B, is a function of wave height. Over a relatively short travel distance $\Delta S \ (\leq 1 \ \text{nm})$, the wave height variation remains small and the damping can be approximated by an exponential decay:

$$H_{n+1} = H_n \exp[-D_B \Delta S], \qquad (1.24)$$

where H_n and H_{n+1} are the wave heights at two locations along wave rays separated by a distance ΔS. Because of nonlinearity, the damping coefficient D_B remains a function of wave height and water depth. The value of D_B corresponding to different soil characteristics defined by the sand diameter D is given in Table 1.7.

Similar calculations have been made recently to account for other dissipative processes. The results are a function of the permeability of the marine sediments, density, sand diameter, bulk modulus, and thickness of sediments.

The implication of these results for EGWWs are:

(1) If a wave originating in deep water is not high enough to break on the continental slope (i.e., $H < 240$ feet, 80 meters) it travels over the long reach of the continental shelf without breaking because the wave damping effect (in addition to dispersion and refraction) overcomes wave shoaling. This is the case, for example, for the East Coast of the US (slope $\approx 10^{-3}$). The wave becomes a spilling breaker only in shallower water and evolves into a fully developed bore where the steepness of the bottom becomes somewhat greater than that of the deeper part of the shelf (Figure 1.18). The steepening of the seafloor occurs in the vicinity of the 30-foot (12 meters) depth contour, which is less than one nautical mile from shore. A series of spilling breakers forms a "surf zone", which extends from the shoreline to the line of breaking inception of the maximum wave. This is known as the "Van Dorn Effect". In the past, the wave damping effect seems to have been underestimated. Consequently, the surf zone was estimated to extend far offshore on the continental shelf. Even though the estimation of the wave dissipation processes are still uncertain, there is nevertheless a general concensus that the Van Dorn effect or the maximum surf zone due to large EGWWs does not extend much beyond that of a very large storm on a long continental shelf. This is due to the importance of the wave damping mechanisms by wave-sea-floor interactions.

(2) If a wave is large enough to break on the continental slope ($H > 240$ feet, 80 meters), it will most likely be a plunging breaker. When the waves reach a long shelf, the breakers will reform as a nonbreaking wave on the continental shelf because of wave damping. (In this process, each breaking

wave could generate 2 or 3 waves of higher frequency, called "solitons".) Then a second breaker will occur nearshore where the bottom becomes steeper. This is not the case for a steep continental shelf such as the West Coast of the US, where once the wave reaches limiting condition and breaks, it will keep traveling towards shore as a spilling breaker or as a fully developed bore (Figure 1.18). This implies that at some slope steepness, a breaking wave will not be damped enough to reform into nonbreaking waves. In reality the phenomenon is ill-defined due to the relative randomness of refractive effects over a long and somewhat irregular bathymetry, allowing breakers traveling on this critical bottom steepness to appear and disappear at random like whitecaps.

14. Wave Breaking and Nearshore Phenomena

Ultimately as the water depth decreases and the wave height increases (by nonlinear shoaling), each wave becomes unstable and, depending on its steepness and the slope of the bottom, will either break as a plunging breaker or as a spilling breaker and surge up onto the shore. As previously stated, this occurs only if the slope of the bottom is steep enough so that the wave damping effect by seafloor interaction does not prevail over wave shoaling.

An approximate simple breaking criterion based on the theory of a limit solitary wave (MacCowan, 1894) has often been used:

$$\frac{H_b}{d_b} = 0.78 \,. \tag{1.25}$$

H_b is the wave height and d_b is the water depth at the breaking inception. More accurately, the family of curves presented in Figure 1.19 which accounts for the effect of bottom slope m and wave steepness should be used (Goda, 1970; Shore Protection Manual, 1977). It is seen, for example, that the limit wave height can actually be larger than the water depth on a steep slope, quite unlike the result obtained using the above approximate criterion. The type of breaker also depends on bottom slope and steepness. An approximate criterion which is valid in the range of the EGWW ($H_b/gT^2 < 5 \times 10^{-3}$) is proposed:

$$m_c \cong \frac{7H_b}{gT^2} \,. \tag{1.26}$$

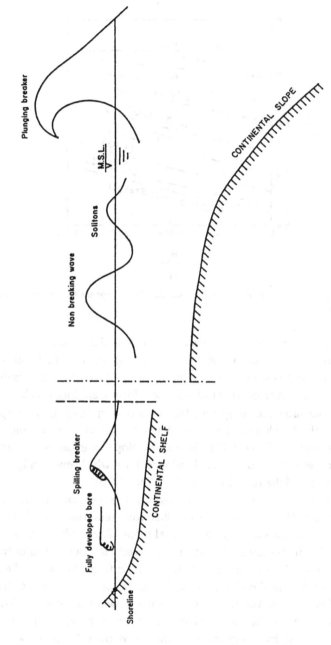

Fig. 1.18. Transformation of a plunging breaker into a nonbreaking wave, and of a spilling breaker into a fully-developed bore nearshore.

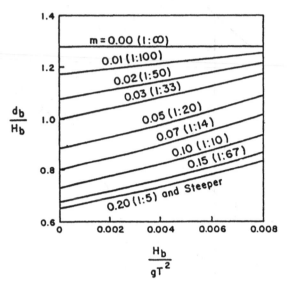

Fig. 1.19. Dimensionless depth of breaking versus breaker steepness for various beach slope m (Goda, 1970).

If $m < m_c$, i.e., on gentle slopes, the wave forms a spilling breaker. If $m > m_c$, a plunging breaker results. A wave which is large enough to break on the continental slope will generally form a plunging breaker, whereas a limit wave which breaks on an inner continental shelf will form a spilling breaker.

Following the breaking inception, the crest of a spilling breaker approximately follows the breaking index curve defined by $H \cong 0.78d$ as long as the slope remains gentle (Figure 1.20). When the slope m exceeds 2×10^{-2} the spilling breaker evolves into a fully developed bore which eventually runs up onto the shore (Le Méhauté, 1963; Divoky *et al.*, 1970).

Whereas the wave energy is dissipated by the free turbulence induced by wave breaking, the wave momentum flux (or "radiation stress" or "wave thrust") needs to be balanced by an external force. The application of the momentum theorem to the area defined by (1) a vertical plane parallel to shore located at wave breaking inception, (2) the seafloor from the line of the breaking inception up to the shoreline, indicates that in order to balance the wave momentum flux towards the shore, an external seaward force is needed. This force is exerted by an additional horizontal pressure component of the seafloor on the fluid, resulting from a gentle rise (the "wave set-up") of the sea level in

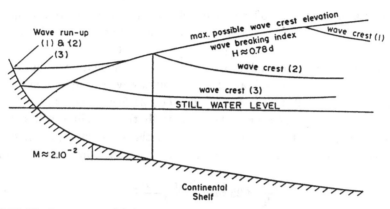

Fig. 1.20. The largest wave (1) does not cause more run-up than the wave (2) which breaks at $M \cong 10^{-2}$. The smaller wave (3) causes less run-up.

the wave breaking zone (the "surf zone") (Figure 1.21). This rise of the mean sea level is, in the case of EGWWs, a non-negligible fraction of the breaking wave height. It is about 10% of the breaking wave height for monochromatic waves, but it can reach a much higher maximum value in the case of transient waves (Le Méhauté, 1971).

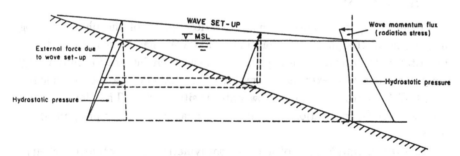

Fig. 1.21. The wave momentum flux is balanced by the increase of bottom pressure due to the wave set-up.

When the wave reaches the shoreline of an open coast, in addition to the wave set-up, it runs up the shore. The maximum possible wave run-up is reached by the wave which breaks at a depth where the bottom slope becomes steeper than about 10^{-2}. Indeed, as previously seen, a larger wave will dissipate its energy farther offshore and will not cause much more run-up than a smaller wave which breaks closer to shore. (The only cause for a larger wave run-up is the wave set-up.) For example, on the East Coast of the US, the continental slope steepens to a value larger than 10^{-2}, by a depth of about 30 feet, which is about 1,000 feet offshore. The maximum wave run-up is roughly that of a wave having 0.78×30 feet in height, i.e., 24 feet. The wave run-up of each wave in the EGWW train is then practically that of a limit solitary wave at that particular water depth. Accordingly, the wave run-up, R_u, on the top of wave set-up can be estimated by the following approximate implicit formula:

$$\frac{R_u}{H_b} \cong 25 \; m = \frac{25(d_b + R_u)}{\ell_H} \, , \qquad (1.27)$$

where H_b is the limit wave height where the bottom slope becomes steeper than $\approx 10^{-2}$, and m is the average slope between that point and the maximum elevation reached by the water edge. ℓ_H is the horizontal distance between the breaking line and the wave run-up.

15. The EGWW Environment on Continental Shelves due to Large Yield Explosion in Deep Water

In the previous sections, a qualitative description of the phenomenology involved in all the phases of evolution of EGWWs has been presented. The quantitative treatment of these phenomena and their integration into a single computer program is extremely complex. The methodology used to calculate the EGWW environment in shallow water is site specific. Therefore, for practical site specific application, a rather sophisticated computer program should be used.

However, a simple case of a plane bathymetry (2D continental shelf) is amenable to a form of presentation which, at least by interpolation, allows a direct approximate assessment of the magnitude of an EGWW environment in a multiplicity of situations of greater complexity.

Many continental shelves and continental slopes can be approximated by a plane bathymetry. This simplifying assumption allows us to classify offshore bathymetries into a few typical categories which can then be used to estimate the EGWW environment. For this reason, a computer program has been developed and applied to a number of typical shelves for a large variety of yields and distances from shore. The corresponding results can then be used for order-of-magnitude site specific applications at any location. In the case of specific complex 3D bathymetries, the results need to be more carefully interpreted due to the vagaries of refractive effects as a function of wave period.

In order to keep the amount of information to an acceptable volume and still provide the reader with the appropriate information for practical use, results have been categorized into typical examples to which other cases could be assimilated.

Three types of shelves have been considered, whose characteristics are given in Table 1.8. Type 1 are the very long shelves ($\approx$ 100 nm, 185 km) exemplified

Table 1.8. Typical offshore 2D bathymetries.

Water depth ft (m)	Distance from shore nm (km)		
	Type 1 very long shelf (100 nm, 185 km)	Type 2 long shelf (60 nm, 111 km)	Type 3 short shelf (6 nm, 11 km)
0 (0)	0 (0)	0 (0)	0 (0)
30 (9.14)	1 (1.85)	1 (1.85)	1 (1.85)
120 (36.6)	50 (92.6)	35 (64.8)	3 (5.55)
300 (91.4)	95 (176)	60 (111)	6 (11.1)
600 (183)	105 (194)	65 (120)	9 (16.6)
3000 (914)	145 (268)	145 (268)	25 (44.3)
6000 (1829)	200 (370)	160 (296)	40 (74.1)
12000 (3657)	230 (425)	190 (372)	50 (92.6)

by the Gulf of Mexico, offshore of Galveston. Type 2 are the long shelves ($\approx$ 60 nm, 111 km) exemplified by South Carolina on the Atlantic coast. Type 3 are the short shelves ($\approx$ 6 nm, 11 km) exemplified by the bathymetry offshore of San Francisoco. It is assumed that the shelves are covered uniformly by a 0.5 mm diameter, uniform sand with a 2.65 specific gravity.

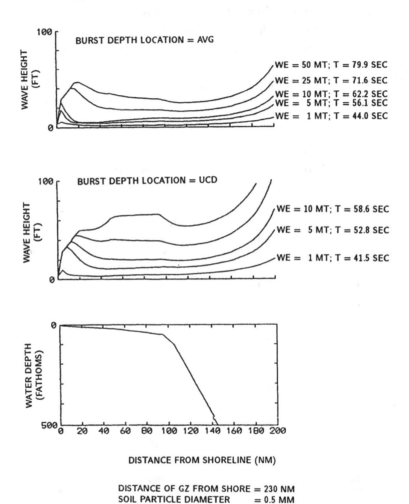

DISTANCE OF GZ FROM SHORE = 230 NM
SOIL PARTICLE DIAMETER   = 0.5 MM

Fig. 1.22. A comparison of wave heights along a perpendicular from GZ to shore for yields 1, 5, 10, 25, 50 MT at UCD and AVG in the case of a very long shelf (Le Méhauté and Soldate, 1986).

The maximum wave heights along a wave ray perpendicular to the shore are presented in a series of 3 nomographs (Figures 1.22–1.24) giving the maximum wave height reached by an EGWW along a line from GZ perpendicular to the bathymetry towards the shore. The period of the maximum wave is also given. The distance is measured from the shore in nm (1,852 meters).

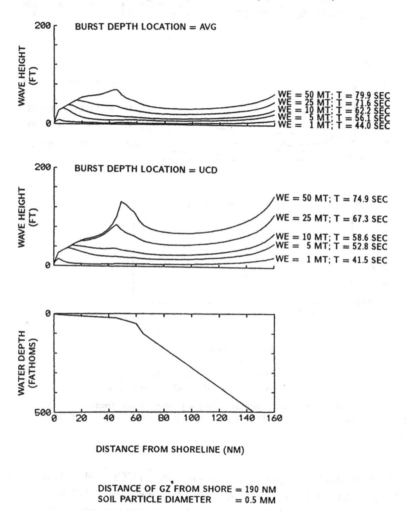

Fig. 1.23. A comparison of wave heights along a perpendicular from GZ to shore for yields 1, 5, 10, 25, and 50 MT at UCD and AVG in the case of a long shelf (Le Méhauté and Soldate, 1986).

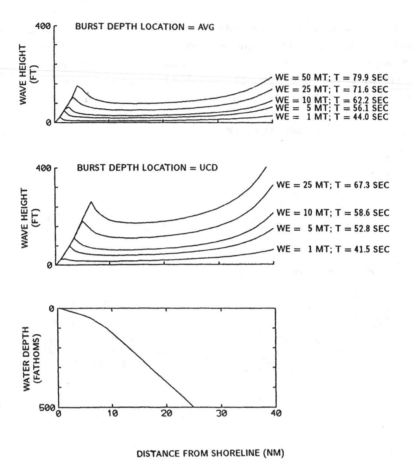

Fig. 1.24. A comparison of wave heights along a perpendicular from GZ to shore for yields 1, 5, 10, 25, and 50 MT at UCD and AVG in the case of a short shelf (Le Méhauté and Soldate, 1986).

There are two figures per shelf corresponding to UCD and AVG respectively. Each figure gives the result corresponding to a distance of GZ from the shore, corresponding grossly to an optimum location: in the case where GZ is closer to shore, the wave generation process is less efficient because of the limits imposed by water depths. In each graph, the yields correspond to 1, 5, 10, 25, and

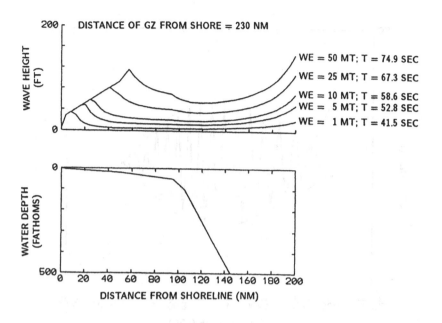

BURST DEPTH LOCATION = UCD
SOIL PARTICLE DIAMETER = 0.0 MM

Fig. 1.25. Frictionless wave propagation to be compared with Figure 1.22 (Le Méhauté and Soldate, 1986).

50 MT respectively, and the periods of the maximum wave corresponding to these yields are also given. In all these figures the maximum extent of the surf zone can be assessed, because the maximum wave height then becomes the image (multiplied by 0.78) of the water depth.

The sensitivity of the shallow water calculations to variations in friction law is illustrated by comparing Figure 1.22 (UCD), obtained with a grain diameter of 0.5 mm, and Figure 1.25 in which friction is completely neglected. It illustrates the importance of friction on long shelves and gives an upper bound to the possible error due to friction. The introduction of energy dissipation by wave-seafloor interaction considerably reduces the extent of the surf zone.

The time history of the free-surface elevation for the first wave train at three locations on a long shelf (Type 2) between GZ and the shore is also given (Figures 1.26–1.28). Note that in the surf zone the wave amplitude is reduced to a constant value proportional to the water depth.

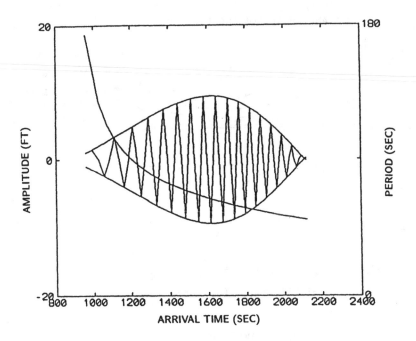

DISTANCE OF GZ FROM SHORE   = 170.0 NM
YIELD STRENGTH   = 10.0 MT
BURST DEPTH LOCATION   = AVG
WATER DEPTH AT BURST   = 8000.0 FT
CRATER RADIUS   = 13258.5 FT
CRATER LIP HEIGHT WRT MSL   = 359.2 FT
PERIOD OF MAXIMUM WAVE   = 62.6 SEC
SOIL PARTICLE DIAMETER   = 0.5 MM

X COORD OBSERVATION POINT   = 243040.0 FT
Y COORD OBSERVATION POINT   = 0.0 FT
DEPTH AT OBSERVATION POINT = 2250.0 FT
DISTANCE FROM GZ TO OBS PT = 50.0 NM

Fig. 1.26. Time history of free-surface elevation of the first wave train at a distance of GZ of 50 nm over a long shelf (Type 2) (Le Méhauté and Soldate, 1986).

16. Accuracy of Results and Research Needs

Even though much progress has been made in the prediction of EGWWs, some uncertainties still remain particularly concerning the wave amplitudes. A lesser uncertainty is associated with the periods of the waves.

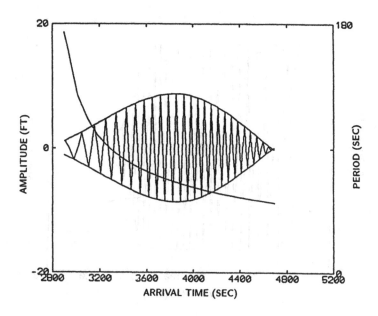

DISTANCE OF GZ FROM SHORE = 170.0 NM
YIELD STRENGTH = 10.0 MT
BURST DEPTH LOCATION = AVG
WATER DEPTH AT BURST = 8000.0 FT
CRATER RADIUS = 13258.5 FT
CRATER LIP HEIGHT WRT MSL = 359.2 FT
PERIOD OF MAXIMUM WAVE = 62.6 SEC
SOIL PARTICLE DIAMETER = 0.5 MM

X COORD OBSERVATION POINT = 607600.0 FT
Y COORD OBSERVATION POINT = 0.0 FT
DEPTH AT OBSERVATION POINT = 300.0 FT
DISTANCE FROM GZ TO OBS PT = 110.0 NM

Fig. 1.27. Time history of free-surface elevation of the first wave train at a distance of GZ of 100 nm over a long shelf (Type 2). Note the increase of the number of wave in the wave train as a result of dispersion (Le Méhauté and Soldate, 1986).

The accuracy of the results depends on the magnitude of errors introduced at each phase of the generation and propagation processes of the assumptions. A large cause of error probably results from extrapolating HE data to the nuclear range. Errors are also due to neglecting nonlinear convective effects: nonlinear refraction, instability of wave profile and apparition of solitons, caustics on 3D bathymetry with crossing of wave orthogonals, reformation

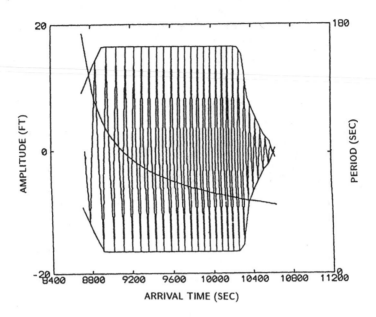

DISTANCE OF GZ FROM SHORE   = 170.0 NM
YIELD STRENGTH   = 10.0 MT
BURST DEPTH LOCATION   = AVG
WATER DEPTH AT BURST   = 8000.0 FT
CRATER RADIUS   = 13258.5 FT
CRATER LIP HEIGHT WRT MSL   = 359.2 FT
PERIOD OF MAXIMUM WAVE   = 62.6 SEC
SOIL PARTICLE DIAMETER   = 0.5 MM

X COORD OBSERVATION POINT   = 929628.0 FT
Y COORD OBSERVATION POINT   = 0.0 FT
DEPTH AT OBSERVATION POINT = 42.3 FT
DISTANCE FROM GZ TO OBS PT = 163.0 NM

Fig. 1.28. Time history of free-surface elevation at a distance of GZ of 163 nm over a long shelf (Type 2). Note the cutoff in wave heights due to depth control (Le Méhauté and Soldate, 1986).

of nonbreaking waves after breaking, transient mass transport, and wave set-up.

Energy dissipation processes by wave-seafloor interaction also involves uncertainty due to both marginal knowledge of the effect of sediments, particularly in the sheet flow regime, soil response, and the lack of precise information on the *in situ* soil characteristics. The comparison of the results with the case where the friction is completely neglected (Figures 1.23 and 1.25 at UCD) established an upperbound for this uncertainty.

It is evident that the research needs have to be dictated by strategic, tactical, or operational implications, as presented in the following section. From a scientific point of view, the research in practically all aspects of the EGWW generation and propagation processes can be pursued for further accuracy. A number of important gaps remain concerning the effects of soil, the intricacies of UCD, the propagation of nonlinear waves over 3D bathymetries, and the dissipative processes due to wave/seafloor interaction.

Despite these limitations, a thorough assessment of the state of the art indicates that at least orders of magnitude are represented. The state of the art is satisfactory in many cases where operational or strategic decisions are to be made. This may not always be sufficient in some important specific cases where the vulnerability of new defense systems is to be considered.

17. Tactical, Strategic, and Operational Implications

Many tactical and strategic implications resulting from the effects of EGWWs have been considered in the past. Overall the effects of EGWWs are mild and generally overcome by other lethal effects. By their very nature of being gravity waves, the accelerations of the wave field are always smaller than the gravity acceleration, g. The only exception is during the growth of the initial cavity. EGWW-induced forces and accelerations are generally small compared to shock waves for example. A structure (for example, a silo built on the continental shelf) can easily be designed to sustain EGWWs. Ships subjected to EGWWs in deep water may be subjected to large amplitude motion, but the wave periods and wave steepness are such that the accelerations do not present more problems than ordinary wind waves or swell, except near GZ. However, EGWWs have a longevity and a range which far exceeds other lethal effects, so that their effects can be felt at long distances from GZ, where other effects have become negligible.

Initially, the effects of EGWWs due to a large yield (> 10 MT) was considered a potentially serious threat. "A tsunami wave of 1 foot in height in deep water may produce 20 feet of run-up; so an EGWW of 100 feet in height in deep water may have devastating effects on the coast and even far inland."

This problem was resolved when it was demonstrated that the analogy between tsunami waves and EGWWs does not hold: EGWWs dissipate most of the energy on the continental shelf, first by wave-seafloor interactions, second in a nonsaturated breaker as previously explained (Le Méhauté, 1963). Therefore, the damages would be limited to coastal installations in much the same as would result from large storm waves. For example, on the East Coast, the sand dunes may be overtopped.

The second concern was about harbor oscillations. Indeed, the periods of EGWWs due to large yields, correspond to the natural period of oscillations of typical harbor basins as seen in Table 1.3. Therefore, even in the case of an explosion at a very large distance away from the entrance of a harbor, an EGW of small amplitude induces a standing wave (seiche) of significant amplitude. This is due to the fact that long waves do not break like wind waves on beaches, rip-rap, or wave absorbers. They reflect their energy back and forth, causing standing wave which at times match the natural oscillations of the harbor basin. Indeed, since the "periods" of EGWWs decrease slowly with time, eventually the excitation at the harbor entrance will match successively the natural periods of oscillation of the basins causing resonance. This phenomenon is known under natural circumstances to create a problem for moored ships whereby their mooring lines tend to break. In the present context of EGWWs, the problem has been investigated in the past, when field tests were considered. Due to the Test Ban Treaty these tests were never done. At the time, it was a source of concern for harbor basins even at a very large distance from GZ. This subject has been amply covered in the past (Le Méhauté, 1971) and will no longer be covered in the present treatise.

The reason that the effect of EGWWs has attracted the most attention is due to the possibility for a large yield to generate a large surf zone on the continental shelf, which indeed could be lethal for transitting surface ships or submarines. For example, a submarine caught in a 100-foot breaker may pitchpole and hit the seafloor violently. An aircraft carrier might capsize. This is known as the "Van Dorn" effect (Moulton and Le Méhauté, 1980). One has seen that this effect was exaggerated. Indeed, in the case of long shelves, as on the East Coast of the US, the wave would be damped by wave-seafloor interaction before reaching the wave limit conditions. In the case of short shelves, as on the West Coast, the surf zone would necessarily be small. Nevertheless, this potential threat still exists. The danger would be particularly acute in the case of a

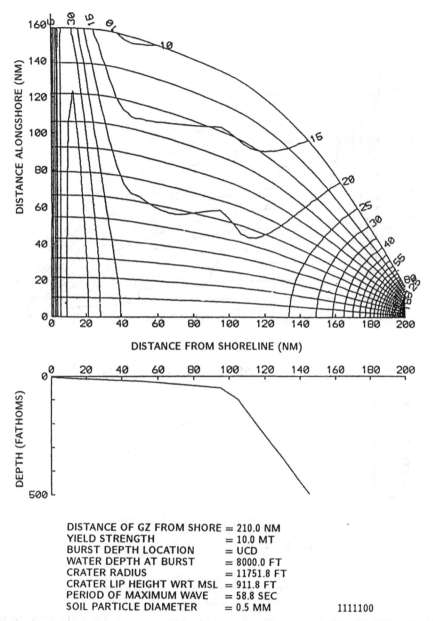

Fig. 1.29. Very long shelf Type 1. Plan view of the maximum wave height (Le Méhauté and Soldate, 1986).

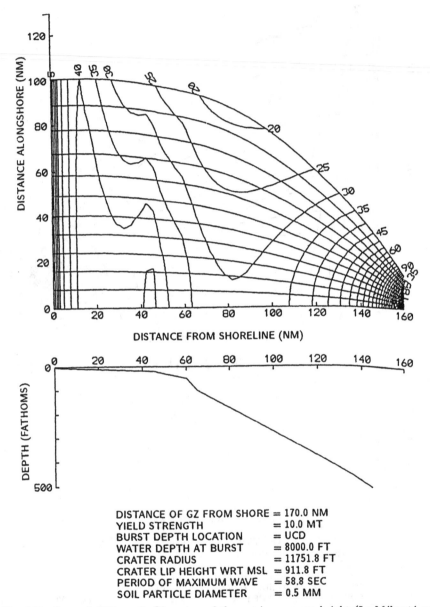

Fig. 1.30. Long shelf Type 2. Plan view of the maximum wave height (Le Méhauté and Soldate, 1986).

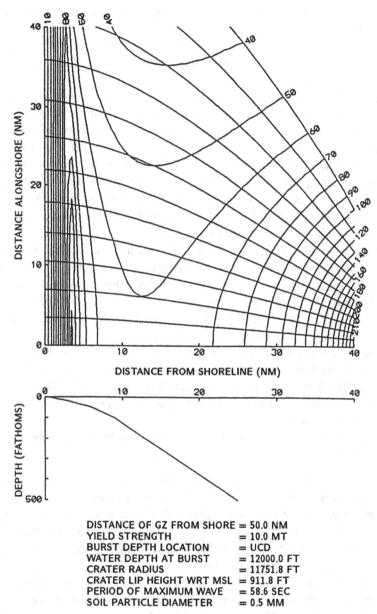

DISTANCE OF GZ FROM SHORE = 50.0 NM
YIELD STRENGTH   = 10.0 MT
BURST DEPTH LOCATION   = UCD
WATER DEPTH AT BURST   = 12000.0 FT
CRATER RADIUS   = 11751.8 FT
CRATER LIP HEIGHT WRT MSL = 911.8 FT
PERIOD OF MAXIMUM WAVE   = 58.6 SEC
SOIL PARTICLE DIAMETER   = 0.5 MM

Fig. 1.31. Short shelf Type 3. Plan view of the maximum wave height (Le Méhauté and Soldate, 1986).

landing operation where an EGWW can cause a considerable amount of casualties to landing ships.

The EGWW environment on the three typical shelves previously described are presented in the plan view in a coordinate system defined by distances offshore X and alongshore Y (Figures 1.29–1.31). This presentation allows assessment of the alongshore lateral spread of the maximum height as well as the direction of the maximum wave. Each graph presents a family of isolines of constant wave height. After breaking, the height of the breaker is assumed to be controlled by the water depth ($H_b/d_b = 0.78$), in which case the isolines of constant height are simply parallel to the bottom contours. The line of maximum breaking inception is defined by the departure from a curve isoline to a straight line defining the maximum extent of the surf zone. For the three nomographs the yield is 10 MT and the distance from GZ is 10 nautical miles from the 6,000-foot (1,829 meters) contour.

Finally, the case of relatively low yields, such as in tactical weapons ($\leq$ 50 KT), has also been investigated in regard to the vulnerability of moored ships and submarines inside a naval base. Submarines at the surface are not very vulnerable to shock waves since they are such hard structures. The question was raised as to whether EGWWs may cause the breaking of mooring lines, the hitting of the seafloor by the submarine hull, or even the grounding of the hull, so that the submarine if not killed, can at least be impaired operationally. It was found that, under certain circumstances, this could occur. In this operation, however, it is realized that the amplitude of EGWWs is depth controlled, and that it is not possible to generate a large EGWW in shallow water, regardless of yield.

For these reasons the optimum general scenario for the best use of effects of EGWWs would be:

(1) A few very large yield detonations offshore of the continental shelf may create a large surf zone on the shelves, possibly causing havoc to a landing fleet, or to transitting surface ships and submarines. However, in order to be significant, the yield must be so large that it may not be available in the present nuclear arsenal (> 50 MT). In this regard, one 50-MT explosion in deep water is more lethal than five 10-MT explosions, in the sense that it would create a larger surf zone. Still, this scenario has limits resulting from energy dissipation processes by wave-seafloor interactions.

(2) In the case of a naval base, a multiplicity of small tactical weapons (<1 KT) would be more efficient than a smaller number of larger ones. This is due

to the fact that the wave heights are depth controlled. The efficiency, ε, of shallow water explosions is given by the ratio of the energy in waves to the energy given by the explosion. It is given by

$$\varepsilon = 0.153 \left(\frac{d}{W^{1/4}} \right)^2, \tag{1.38}$$

where the water depth d is in feet, and the yield W in pounds of TNT (Le Méhauté and Khangaonkar, 1991). Typically, at most only 4% of the explosion energy is transmitted in the form of water waves.

Underwater explosions are poor wave generators. For these reasons the subject presented in this treatise has traditionally taken a back seat in the research on the effects of explosion. Nevertheless, the EGWW needs to be evaluated and known. For example, the effect of EGWWs was used erroneously as a reason for not locating launching silos on the continental shelf. Whatever the many reasons for such a decision, the EGWW should not have been one, as one can certainly design an underwater structure capable of withstanding their impact. EGWW-induced pressure has also been considered for triggering pressure mines on beaches during Operation Desert Storm. It was called Operation "Storm Wave". It evidently did not work as the reader of this treatise might have surmised. Whatever its importance, the subject of EGWWs belongs to the arsenal of effects which need to be known.

Chapter 2

LINEAR THEORY OF IMPULSIVELY
GENERATED WATER WAVES ON
A HORIZONTAL BOTTOM

1. Derivation of a General Mathematical Model

The theoretical problem consists of determining the wave field resulting from a localized disturbance of the free surface. The corresponding mathematical model is a synthesized approach of the methods presented by Lamb (1932), Kranzer and Keller (1959), and Stoker (1957). The method developed here is the most general and is able to accept as initial conditions any form of initial disturbances (Le Méhauté et al., 1987).

The wave motion is defined with respect to time t and space (r, z, θ) in a cylindrical coordinate system centered at ground zero (GZ). The parameter r is the radial distance from GZ and z the vertical distance (positive upward) from the still-water level (SWL). The water depth d is constant and the motion is symmetrical with respect to the axis oz ($\frac{\partial}{\partial \theta} = 0$). The motion is irrotational allowing the definition of a potential function $\phi = \phi(r, z, t)$ and the free surface is $\eta(r, t)$.

The motion is nondimensionalized with respect to water depth d, such that the dimensionless water depth becomes unity and

$$r^*,\, z^*,\, \eta^* = \frac{(r,\, z,\, \eta)}{d}\,,$$

$$t^* = t\sqrt{\frac{g}{d}}\,,$$

$$V^*(u_r,\, w) = \frac{V(u_r,\, w)}{\sqrt{gd}}\,,$$

$$p^* = \frac{p}{\rho gd}\,,$$

$$\sigma^* = \sigma\sqrt{\frac{d}{g}}\,,$$

$$\phi^* = \frac{\phi}{d\sqrt{gd}}\,,$$

$$(2.1)$$

where g is the gravity acceleration, V is the velocity, u_r and w denote the radial and vertical particle velocity, respectively, p the pressure, and σ frequency.

In terms of dimensionless variables, the equation of motion is given by

$$\nabla^2\phi = \phi_{zz} + \phi_{rr} + \frac{1}{r}\phi_r\,. \tag{2.2}$$

It is understood that the notation "*" has been dropped from the nondimensional variables and the equation is dimensionless. In the following, all notations are dimensionless unless otherwise specified. Also, the nonlinear terms are neglected. Accordingly, the free-surface boundary conditions are

$$\phi_z = \eta_t \quad \text{for} \quad z = \eta \cong 0, \quad \text{and}$$

$$\phi_t = -\eta - p\,. \tag{2.3}$$

At the bottom,

$$\phi_z = 0 \quad \text{for} \quad z = -1\,. \tag{2.4}$$

A general solution to Equation (2.2) is of the form

$$\phi = F(r)G(z)[A\sin\sigma t + B\cos\sigma t]\,. \tag{2.5}$$

Inserting (2.5) into (2.2) yields

$$\frac{G''}{G} = \frac{1}{\cdot F}\left(F'' + \frac{1}{r}F'\right) = k^2, \tag{2.6}$$

where k is an arbitrary constant and the primes denote partial differentiation with respect to the argument. The solutions to Equation (2.6) are

$$F(r) = A_1 J_0(kr) + B_1 Y_0(kr) \tag{2.7}$$

$$G(z) = A_2 e^{kz} + B_2 e^{-kz}. \tag{2.8}$$

where J_0 and Y_0 are zero order Bessel functions of the first and second kind, respectively. Since the solution must be finite when $r \to 0$

$$F(r) = A_1 J_0(kr) \tag{2.9}$$

and inserting Equation (2.4) into Equation (2.8) gives straightforwardly

$$G(z) = \cosh k(1 + z). \tag{2.10}$$

Then the solution is of the form

$$\phi = J_0(kr)\frac{\cosh k(1+z)}{\cosh k}[A \sin \sigma t + B \cos \sigma t], \tag{2.11}$$

which, when introduced into Equation (2.3), yields the dispersion relationship

$$\sigma^2 = k \tanh k; \tag{2.12}$$

k is the dimensionless wave number ($k = 2\pi d/L$). Since the solution is linear and k is arbitrary, the most general solution is given by a continuous linear superposition of the form of the solution presented in Equation (2.11), i.e.,

$$\phi(r, z, t) = \int_0^\infty J_0(kr)\frac{\cosh k(1+z)}{\cosh k}[A(k) \sin \sigma t + B(k) \cos \sigma t]\, k\, dk, \tag{2.13}$$

The coefficients $A(k)$ and $B(k)$ are defined from the initial free-surface conditions at time $t = 0$.

2. Initial Conditions

The problem now consists of defining the initial conditions imposed on the fluid by the underwater explosion. In hydrodynamic terms, this could be done by considering that the explosion causes either:

(1) a positive or negative instantaneous pressure or impulse on the free surface $I(r_0, 0)$,
(2) an initial free-surface negative displacement in the form of a cavity or crater, or a positive elevation forming a dome above the still-water level $\eta_0(r_0, 0)$, or
(3) an initial downward or upward velocity of the free surface $w_s(r_0, 0)$.

The problem now consists of relating these physical definitions to the potential function (Equation 2.13) at time $t = 0$.

(1) An initial dimensionless impulse I at the surface is defined by

$$p(r_0) = I(r_0)\delta(t) \qquad \text{at} \quad z = \eta_0 = 0\,, \tag{2.14}$$

where δ is the Dirac delta function which is 1 at time $t = 0$ and is 0 otherwise. Inserting this into Equation (2.3) and since

$$\int p\,dt = \int I(r_0)\delta(t)\,dt = I(r_0) = \int_{t=0} \phi_t\,dt = \phi(r_0, 0)\,, \tag{2.15}$$

then

$$I(r_0, 0) = \phi(r_0, 0)\,. \tag{2.16}$$

(2) Also referring to Equation (2.3) where $p = 0$, the initial dimensionless disturbance of the free-surface elevation $\eta_0(r_0, 0)$ in terms of the potential function is:

$$\eta_0(r_0, 0) = -\phi_t(r_0, 0)\,. \tag{2.17}$$

(3) If the initial conditions are defined by a dimensionless velocity of the free surface w_s, then

$$w_s(r_0, 0) = \phi_z(r_0, 0)\,, \tag{2.18}$$

or alternately,

$$w_s(r_0, 0) = -\phi_{tt}(r_0, 0)\,. \tag{2.19}$$

Then Equation (2.13) at time $t = 0$ reduces to one of the following, depending on the type of initial condition which is used.

$$I(r_0) = \int_0^\infty dk\, kJ_0(kr)\, B(k)\,, \tag{2.20}$$

$$\eta_0(r_0) = -\int_0^\infty dk\, kJ_0(kr)\, \sigma(k)\, A(k)\,, \tag{2.21}$$

$$w_s(r_0) = \int_0^\infty dk\, kJ_0(kr)\, \sigma^2(k)\, B(k)\,. \tag{2.22}$$

Equations (2.20)–(2.22) can be inverted by virtue of the Fourier-Bessel Theorem. Then introducing the notation H_I, H_η, H_w one finds:

$$H_I(k) = B(k) = \int_0^R I_0(r_0)J_0(kr_0)r_0 dr_0\,, \tag{2.23}$$

$$H_\eta(k) = -A(k)\,\sigma(k) = \int_0^R \eta_0(r_0)J_0(kr_0)r_0 dr_0\,, \tag{2.24}$$

and

$$H_w(k) = B(k)\,\sigma^2(k) = \int_0^R w_s(r_0)J_0(kr_0)r_0 dr_0\,, \tag{2.25}$$

where R is the dimensionless maximum value of r_0 corresponding to the limit of the initial disturbances. (The integrals from R to infinity are nil.) Note that the H_is are the Hankel transforms of I, η_0, w_s respectively. They are independent of the physical definition of the initial disturbance, i.e., for an identical mathematical definition of I, η_0, and w_s, the H_is are mathematically the same but they are dimensionally different if converted into dimensional form.

Inserting Equations (2.23)–(2.25) into Equation (2.13) yields (in the three cases considered here above):

$$\phi_I(r,\,z,\,t) = \int_0^\infty dk\, H_I(k)J(k)\, \cos\sigma t\,, \tag{2.26}$$

$$\phi_\eta(r,\,z,\,t) = -\int_0^\infty dk\, H_\eta(k)J(k)\, \frac{\sin\sigma t}{\sigma}\,, \tag{2.27}$$

$$\phi_w(r,\,z,\,t) = \int_0^\infty dk\, H_w(k)J(k)\, \frac{\cos\sigma t}{\sigma^2}\,, \tag{2.28}$$

where

$$J(k) = k J_0(kr) \frac{\cosh k(1+z)}{\cosh k} \,. \tag{2.29}$$

The free-surface elevation given by Equation (2.3) then becomes in the three cases:

$$\eta_I(r,\,t) = \int_0^\infty k\,dk\,H_I(k) J_0(kr)\,\sigma\,\sin\sigma t\,, \tag{2.30}$$

$$\eta_\eta(r,\,t) = \int_0^\infty k\,dk\,H_\eta(k) J_0(kr)\,\cos\sigma t\,, \tag{2.31}$$

$$\eta_w(r,\,t) = \int_0^\infty k\,dk\,H_w(k) J_0(kr)\,\frac{\sin\sigma t}{\sigma} \,. \tag{2.32}$$

Note that the only differences in the time functions between the three forms of initial conditions are

$$f(t) = \sigma\,\sin\sigma t,\ \cos\sigma t,\ \frac{\sin\sigma t}{\sigma}\,, \tag{2.33}$$

in Equations (2.30)–(2.32) respectively. Note also that

$$\eta_\eta = \frac{\partial\eta_w}{\partial t}\,, \tag{2.34}$$

and

$$\eta_I = -\frac{\partial\eta_\eta}{\partial t} = -\frac{\partial^2\eta_w}{\partial t^2} \,. \tag{2.35}$$

The formulation (2.30)–(2.32) is very general and is valid for any kind of initial conditions and their linear superposition so that the general solution is

$$\eta(r,\,t) = \int_0^\infty J_0(kr)[H_\eta(k)\,\cos\sigma t$$

$$+\,(H_I(k)\,\sigma + H_w(k)\,\sigma^{-1})\,\sin\sigma t]\,k\,dk\,. \tag{2.36}$$

In the general case, the integral (2.36) cannot be solved analytically. However, there exists a number of numerical methods which allows its integration. One is by Fast Fourier transform. There also exists a number of simplifying assumptions which can be introduced, which permits its analytical integration.

3. Integration by Fourier Transform

Consider the time history of the free-surface elevation $\eta(t)$ at a particular distance r from GZ. Then consider the Fourier transform pair defined by

$$\eta(t) = \frac{1}{2\pi} \int_{-\infty}^{\infty} A(\sigma)e^{i\sigma t}d\sigma, \tag{2.37}$$

and

$$A(\sigma) = \int_{-\infty}^{\infty} \eta(t)e^{-i\sigma t}dt, \tag{2.38}$$

where $A(\sigma)$ denotes the complex Fourier coefficients defined by

$$A(\sigma) = F_1(\sigma) + iF_2(\sigma). \tag{2.39}$$

$F_1(\sigma)$ is the real part and $F_2(\sigma)$ is the imaginary part of $A(\sigma)$. Introducing Equation (2.39) into Equation (2.37)

$$\eta(t) = \frac{1}{2\pi} \int_{-\infty}^{\infty} [F_1(\sigma)e^{i\sigma t}d\sigma + iF_2(\sigma)e^{i\sigma t}d\sigma], \tag{2.40}$$

which is

$$\eta(t) = \frac{1}{2\pi} \int_{-\infty}^{\infty} [F_1(\sigma)\cos(\sigma t) - F_2(\sigma)\sin(\sigma t)]d\sigma$$

$$+ \frac{i}{2\pi} \int_{-\infty}^{\infty} [F_2(\sigma)\cos(\sigma t) + F_1(\sigma)\sin(\sigma t)]d\sigma. \tag{2.41}$$

Since $\eta(t)$ is real, the Fourier coefficients follow the symmetry defined by $A(-\sigma) = A^*(\sigma)$ where A^* is the complex conjugate of A. This implies that

(1) $F_1(\sigma)$ is an even function, or $F_1(-\sigma) = F_1(\sigma)$; $\qquad$ (2.42)

and

(2) $F_2(\sigma)$ is an odd function, or $F_2(-\sigma) = -F_2(\sigma)$. $\qquad$ (2.43)

Therefore, the second part of Equation (2.41)

$$\frac{i}{2\pi} \int_{-\infty}^{\infty} [F_2(\sigma)\cos(\sigma t) + F_1(\sigma)\sin(\sigma t)]d\sigma = 0. \tag{2.44}$$

Thus,

$$\eta(t) = \frac{1}{2\pi} \int_{-\infty}^{\infty} [F_1(\sigma) \cos(\sigma t) - F_2(\sigma) \sin(\sigma t)] d\sigma \,, \tag{2.45}$$

which can be written using Equations (2.42) and (2.43)

$$\eta(t) = \frac{1}{\pi} \int_{0}^{\infty} [F_1(\sigma) \cos(\sigma t) - F_2(\sigma) \sin(\sigma t)] d\sigma \,. \tag{2.46}$$

This is true for a time series recorded at any distance r, and since the group velocity

$$V(k) = \frac{d\sigma}{dk} \,, \tag{2.47}$$

then Equation (2.46) becomes

$$\eta(r, t) = \frac{1}{\pi} \int_{0}^{\infty} [F_1(\sigma) \cos(\sigma t) - F_2(\sigma) \sin(\sigma t)] V(k) \, dk \,. \tag{2.48}$$

Equations (2.36) and (2.48) are identical provided

$$F_1(\sigma) = \frac{\pi k J_0(kr)}{V(k)} H_\eta(k) \,, \tag{2.49}$$

and

$$F_2(\sigma) = \frac{\pi k J_0(kr)}{V(k)} [H_I(k)\sigma + H_w(k)\sigma^{-1}] \,. \tag{2.50}$$

Thus, if the initial Hankel transforms $H(k)$ are known, the solution is simply Equation (2.37), where the real and imaginary parts of $A(\sigma)$ [Equation (2.39)] are given by Equations (2.49) and (2.50).

4. Overview of Analytical and Numerical Solutions

In the case where the initial conditions I, η_0, w_s are defined numerically, then the problem is solved numerically. First, the $H(k)$ functions are determined [Equations (2.23)–(2.25)]. In general, this numerical integration does not present much difficulty. For r_0 small, $J_0(kr_0)$ can be replaced by the first few terms of its power series expansion.

The numerical integration of Equation (2.36) is much more complex since it has to be carried out from 0 to infinity, and the integrand is, in general, a product of slowly converging oscillating functions. Even in the cases where the initial disturbances allow an analytical solution to the Hankel transforms [Equations (2.23)–(2.25)], an exact analytical solution to Equation (2.36) is not possible. This is due to the complexity introduced by the dispersion relationship [Equation (2.12)] through the cosine and sine functions of the integrand. At this time the standard procedure is by the Fast Fourier transform, such as given by Equation (2.37) combined with Equation (2.49) and/or (2.50).

However, a number of simplifying assumptions can be introduced in the integration of Equation (2.36). Then partially valid solutions are obtained when combined with some analytical forms of the $H(k)$. The solutions are obtained in closed forms, from which a clearer picture of impulsive-generated water waves emerges.

In order to understand the relative validity of the simplifying assumptions, one must note that the energy spectrum is distributed as a function of the wave number k. The small values of k correspond to the wave elements which travel the fastest (i.e., the leading waves), the larger values of k should yield the wave trailing behind. Accordingly, any kind of assumption which relates to the k values is indicative of the relative accuracy which is obtained for different parts of the wave train. The relative validity of the simplifying assumptions is dictated by the location of the energy peaks within the wave train and their relative effects.

For example, if the energy is mostly in the leading waves, then an assumption based on $k \ll 1$ would be valid. If the trailing waves are the most important, then assumptions based on $k \gg 1$ are introduced. In general, we may expect that the energy spectrum will peak near a wave number corresponding to a wave length of a small multiple of the initial disturbance radius.

Referring to Equation (2.12), it appears immediately that the assumption $k \ll 1$ leads to $\sigma = k$. Then, neglecting that part of the integral (2.39) from a large value of k to infinity should yield a negligible error. In principle, this assumption is valid for the tip of the leading wave in very shallow water, or for large values of the parameter R ($R \gg 1$). The solution is then nondispersive.

In the case where $k < 1$:

$$\sigma = [k \tanh k]^{1/2} \cong k - \frac{k^3}{6} , \qquad (2.51)$$

and a mildly dispersive solution is obtained, which is valid for the leading wave.

When $k \gg 1$, then

$$\sigma = k^{1/2} . \qquad (2.52)$$

The leading wave solution is not valid, but the trailing wave, in deep water, for small values of R is then obtained.

The solution at GZ ($r = 0$) or for small values of r is also of great interest, since it allows the determination of the time history of the dome or cavity caused by the explosion. Then, when $r = 0$, i.e., at GZ:

$$J_0(kr) = 1 \tag{2.53}$$

which allows a considerable simplification (both in the analytical and numerical schemes). A number of nondispersive and dispersive solutions can then be obtained. Actually, dispersion needs time and distance to be effective. Therefore, the dispersive ($\sigma = k^{1/2}$) and nondispersive ($\sigma = k$) solutions are nearly identical quantitatively at GZ. (It will be seen that the motion at GZ is nearly nonoscillatory and stops very rapidly.)

Probably the most frequent approximation is $kr \gg 1$ which is valid for the trailing wave at a large distance from GZ. Then, $J_0(kr)$ can be replaced by the first term of its asymptotic expansion, namely,

$$J_0(kr) = \left(\frac{2}{\pi kr}\right)^{1/2} \cos\left(kr - \frac{\pi}{4}\right). \tag{2.54}$$

The insertion of this approximation into Equation (2.39) allows us to define a number of analytical solutions which are obtained by the method of the stationary phase. This method is described in the following section. It is emphasized that this solution is not valid for any waves close to GZ. Despite these limitations, it has been the most widely used method in the field of EGWWs.

Finally, one can also mention the power series solutions which are valid for all wave numbers that $k < \pi/2$. The method is based on the combination of the power series in k, i.e., each function of the integrand in Equation (2.39) is replaced by a power series. The terms of equal value in k are combined and integrated analytically. This method was initiated by Whalin (1965) and pursued to a power k^9 by Le Méhauté et al. (1987). This method has now been dropped in favor of the others and will not be presented here.

In summary one has

(1) the numerical solutions uniformly valid for all values of k and r obtained by the Fast Fourier transform (Le Méhauté et al., 1989; Wang et al., 1989; Khangaonkar, 1990).

(2) the nondispersive solutions valid for the tip of the leading wave and nearly valid within the confines of the initial disturbance (Wang and Le Méhauté 1987; Le Méhauté *et al.* 1987)

(3) the mildly dispersive solution valid for the leading wave (Wang, 1987), and

(4) the asymptotic solutions and the method of stationary phase valid for the trailing wave at a distance from GZ (Kranzer and Keller, 1959; Le Méhauté, 1971). The relative limits of these approaches will be presented in Section 2.10.

5. Analytical Solutions — The Stationary Phase Approximation

The method of stationary phase was originally developed by Kelvin (see Jeffreys and Jeffreys, 1956, for example).

Assuming $kr \gg 1$, i.e., at a large distance from GZ, $J_0(kr)$ in Equation (2.36) is replaced by its asymptotic form (Equation 2.54). Then Equation (2.36) can be written as:

$$
\eta(r,\, t) = \int_{-\infty}^{\infty} \left\{ F_1(k) \cos \left(\sigma t - kr + \frac{\pi}{4} \right) - F_2(k) \sin \left(\sigma t - kr + \frac{\pi}{4} \right) \right\}
$$

$$
\times \left(\frac{k}{\pi r} \right)^{1/2} dk, \tag{2.55}
$$

where

$$
F_1(k) = H_\eta(k) \tag{2.56}
$$

$$
F_2(k) = H_I(k)\sigma + H_w(k)\sigma^{-1}. \tag{2.57}
$$

For a fixed value of r, the solution appears as the sum of sinusoidal functions varying rapidly with time, modulated in amplitude by the strength

$$
F(k) \left(\frac{k}{\pi r} \right)^{1/2}.
$$

Over the range of integration in k, the contributions of cosine and sine tend to cancel for large r and t. The strength depends only on the particular mode $k = k_i$ which arrives at that time (for which $F(k) = F(k_i)$) and the neighborhood points around k_i.

Since, energy travels at group velocity $V(k)$, each mode k_i will exit at time t such as

$$V(k_i) = \frac{d\sigma}{dk} = \frac{r}{t} . \tag{2.58}$$

which is equivalent to say that the phase

$$\psi(k) = \left(\sigma - \frac{kr}{t}\right) , \tag{2.59}$$

is constant. Indeed

$$\frac{\partial \psi(k)}{dk} = \left[\frac{d\sigma}{dk} - \frac{r}{t}\right] = 0 , \tag{2.60}$$

which corresponds to a stationary phase in a small neighborhood around k_i, where $\psi_i = \psi(k_i)$, and

$$\psi(k) \cong \psi_i + \frac{1}{2}(k - k_i)^2 \psi_i'' + \cdots . \tag{2.61}$$

For simplicity, consider the term $F_1(k)$ only. By taking the contributions of the wave elements arriving at large r and t according to Equation (2.58), then Equation (2.55) becomes

$$\eta(r, t) = \left(\frac{k_i}{\pi r}\right)^{1/2} F_1(k_i) Re \int_{-\infty}^{\infty} \exp\left[it\left(\psi_i + \frac{1}{2}(k - k_i)^2 \psi_i''\right)\right] dk , \tag{2.62}$$

or

$$\eta(r, t) = \left(\frac{k_i}{\pi r}\right)^{1/2} F_1(k_i) Re \int_{-\infty}^{\infty} \exp[i\psi_i t] \exp\left\{\pm i\left[\left(\frac{1}{2}|\psi_i'' t|\right)^{1/2}(k - k_i)\right]^2\right\} dk , \tag{2.63}$$

where the $\pm$ sign is the same as the sign of ψ_i''. It follows that

$$\eta(r, t) = \left(\frac{k_i}{\pi r}\right)^{1/2} F_1(k_i) \left[\frac{2\pi}{t\psi_i''}\right]^{1/2} Re\left\{\exp\left[i\left(\psi_i t \pm \frac{\pi}{4}\right)\right]\right\} . \tag{2.64}$$

Note

$$\psi_i'' = \frac{dV(k_i)}{dk} . \tag{2.65}$$

Therefore, referring to Equation (2.59), by considering all the stationary points for which $k = k_i$, the solution becomes

$$\eta(r,\, t) = \frac{F_1(k)}{r} \left(\frac{kV(k)}{-\dfrac{dV}{dk}} \right)^{1/2} \cos(\sigma t - kr)\,. \qquad (2.66)$$

Similar calculations done on $F_2(k)$ yields a similar expression. Replacing the $F(k)$ function by Equations (2.56) and (2.57) and adding together gives the general analytical solution:

$$\eta(r,\, t) = \frac{1}{r} \left(\frac{kV(k)}{-\dfrac{dV}{dk}} \right)^{1/2} L(k)\,, \qquad (2.67)$$

where

$$L(k) = H_\eta(k) \cos(\sigma t - kr) + (H_I(k)\sigma + H_w(k)\sigma^{-1}) \sin(\sigma t - kr)\,.$$

The group velocity $V(k)$ is given by differentiating the dispersion relationship (2.12)

$$V(k) = \frac{d\sigma}{dk} = \frac{r}{t} = \frac{1}{2}\left[\frac{\tanh k}{k}\right]^{1/2} + \frac{1}{2\cosh^2 k}\left[\frac{k}{\tanh k}\right]^{1/2}\,. \qquad (2.68)$$

This is equivalent to considering that the wave motion is defined by rapidly time varying functions such as

$$f(r,\, t) = \left\{ {\cos \atop \sin} [(k \tanh k)^{1/2}t - kr] \right\}\,, \qquad (2.69)$$

where the amplitudes of the modulations are

$$\left.\begin{array}{c} A_\eta(k) \\ A_{I,w}(k) \end{array}\right\} = \frac{1}{r}\left[\frac{kV(k)}{-\dfrac{dV}{dk}} \right]^{1/2} \left\{ \begin{array}{c} H_\eta(k) \\ H_I(k)\sigma + H_w(k)\sigma^{-1} \end{array} \right\}\,. \qquad (2.70)(2.71)$$

For small values of r, the stationary phase approximation is no longer valid, because the asymptotic value of the $J_0(k, r)$ given by Equation (2.54) is no longer valid.

6. Determination of the Numerical Initial Disturbances

The initial conditions I, η_0, w_s are fictitious linear equivalents to the complex explosion wave generation process. Therefore, it will be illusory to determine these from the explosion physics and bubble dynamics within the confines of the linear approximation. In particular, since the energy dissipation processes are neglected, one should expect that the real water motion will be larger than its linear equivalent. Therefore, the only process by which the initial conditions can be determined and related to explosion parameters is through the analysis of wave records.

Given the time history $\eta(t)$ at a distance r from GZ, and referring to Equations (2.38) and (2.39), then

$$F_1(\sigma) = Re\, A(\sigma)\,, \tag{2.72}$$

and

$$F_2(\sigma) = Im\, A(\sigma)\,, \tag{2.73}$$

which are obtained from the forward Fourier transform of the wave record $\eta(t)$ (Le Méhauté *et al.*, 1989; Wang *et al.*, 1989; Khangaonkar and Le Méhauté, 1991).

Given $F_1(\sigma)$, the function $H_\eta(k)$ can be obtained from Equation (2.49):

$$H_\eta(k) = \frac{F_1(\sigma)V(k)}{\pi k J_0(kr)}\,. \tag{2.74}$$

Theoretically, as long as the wave propagation follows the linear wave theory, all the wave records $\eta(t)$, obtained at different distances r for the same experiments, should yield a unique function distribution for $H_\eta(k)$. Then by applying the Fourier-Bessel theorem and inverting Equation (2.24), the initial condition for $\eta_0(r_0)$ is obtained:

$$\eta_0(r_0) = \int_0^\infty H_\eta(k) J_0(kr_0) k\, dk\,. \tag{2.75}$$

The numerical problems associated with that procedure are resolved by replacing F_1/J_0 in Equation (2.74) by the equation of the envelope $J_0 F_1/(J_0^2 + \varepsilon)$ (ε is a very small number).

Now given $F_2(\sigma)$, by the same procedure one obtains:

$$H_I(k)\sigma + H_w(k)\sigma^{-1} = \frac{F_2(\sigma)V(k)}{\pi k J_0(kr)} . \qquad (2.76)$$

The solution for $H_I(k)$ and $H_\eta(k)$ cannot be resolved without additional information. The simplest method consists of assuming that the initial condition is either that of a free-surface elevation $\eta_0(r_0)$ and an impulse $I(r_0)$ with the effect (contribution) of $w_s(r_0)$ already included into the two others, or it is that of $\eta_0(r_0)$ and a velocity $w_s(r_0)$ with the effect of an initial impulse included in these two.

In the first case, inverting Equation (2.23) and inserting the expression for $H_I(k)$ gives

$$I(r_0) = \int_0^\infty \frac{F_2(\sigma)V(k)}{\sigma \pi J_0(kr)} J_0(kr_0)dk , \qquad (2.77)$$

and in the second case, inverting Equation (2.25) and inserting $H_w(k)$ results in

$$w_s(r_0) = \int_0^\infty \frac{\sigma F_2(\sigma)V(k)}{\pi J_0(kr)} J_0(kr_0)dk . \qquad (2.78)$$

Then $\eta_0(r_0)$ and $I(r_0)$ or $\eta_0(r_0)$ and $w_s(r_0)$ are related to the explosion parameters of the corresponding wave records.

7. Analytical Forms of Initial Disturbances, Closed Form Solutions

In order to parameterize the initial disturbances as a function of the explosion parameters, it is convenient to match the curves I, η_0, and $w_s(r_0)$ obtained numerically by Equations (2.75), (2.77), and (2.78) with analytical functions which can be Hankel transformed. Then the EGWW model will be obtained by simply relating the corresponding function parameters to the explosion parameters.

Referring to Equations (2.23), (2.24), and (2.25), note that the Hankel transforms H_I, H_η, H_w are identical for the three physical definitions (impulse, free-surface elevation, and velocity) of the initial conditions if I, η_0, and w_s are defined by the same mathematical function, regardless of their physical meaning.

There exists a number of functions for I, η_0, w_s which have been Hankel transforms and are physically reasonable candidates for fictictious linear representation of explosion-generated disturbance. Some of these candidate functions can be indiscriminately positive or negative in which case, the corresponding sign for $\eta(t)$ is changed from positive to negative.

Seven solutions are presented in Table 2.1. These solutions could be considered indiscriminately either as an impulse, a free-surface elevation, or a velocity, so that there are in fact 21 types of candidate solutions. Furthermore, since the theory is linear, these 21 solutions can be added linearly with any appropriate coefficients. (The maximum amplitude of the initial disturbance displayed in Table 2.1 is unity.) Accordingly, the most adaptable mathematical linear models for EGWWs can be represented by any of the 21 functions of the following forms and their linear superposition.

$$
\begin{matrix}
\eta_I(r,\,t) \\
\eta_\eta(r,\,t) = \pm \displaystyle\int_0^\infty \\
\eta_w(r,\,t)
\end{matrix}
\left|\begin{matrix} A_I \\ A_\eta \\ A_w \end{matrix}\right|
\left|\begin{matrix}
\dfrac{R}{k} J_1(kR) \\[2mm]
\dfrac{2}{k^2} J_2(kR) \\[2mm]
\dfrac{R}{k} J_3(kR) \\[2mm]
\dfrac{4}{k^2} J_4(kR) \\[2mm]
R^2 e^{-kR} \\[2mm]
\dfrac{R^2}{2} \exp\left[-\left(\dfrac{kR}{2}\right)^2\right] \\[2mm]
(kR)^{n-1} R^2 \exp(-kR)
\end{matrix}\right|
\left|\begin{matrix}
\dfrac{\sin \sigma t}{\sigma} \\[2mm]
\cos \sigma t \\[2mm]
\sigma \sin \sigma t
\end{matrix}\right| J_0(kr)k\,dk \qquad (2.79)
$$

Using the stationary phase approximation, the corresponding $\eta(r,\,t)$ functions can be expressed in closed form as follows

$$
\left.\begin{array}{c}
\eta_I(r, t) \\
\eta_\eta(r, t) \\
\eta_w(r, t)
\end{array}\right\}
\left.=\begin{array}{c}
A_i \\
A_\eta \\
A_w
\end{array}\right\}
H(kR)\frac{1}{r}\left(\frac{kV(k)}{-\dfrac{dV}{dk}}\right)^{1/2}
\left\{\begin{array}{l}
\dfrac{1}{\sigma}\sin(kr - \sigma t) \\[4pt]
\cos(kr - \sigma t) \\[4pt]
\sigma\sin(kr - \sigma t)
\end{array}\right.
\qquad (2.80)
$$

where $H(kR)$ is the Hankel transform of the initial conditions, regardless of their physical characteristics.

Note that some solutions (Cases 3 and 4) satisfy continuity when the free-surface disturbances are defined by a free-surface elevation $\eta_0(r_0)$. For the other cases (Cases 1, 2, and 5), the displaced volume is determined by

$$
\text{Vol} = 2\pi \int_0^R \eta(r)r\,dr . \qquad (2.81)
$$

Also, the corresponding potential energy E is determined,

$$
E = \rho g 2\pi \int_0^R \left[\int_0^{\eta_0} \eta(r)d\eta \right] r\,dr . \qquad (2.82)
$$

The theoretical parameters are the A and R coefficients. Note also that the last Hankel transform (Case 7) has an additional parameter, n.

Considering the number of possibilities, it is easily realized that practically any kind of numerically determined initial distubances, Equations (2.75), (2.77), and (2.78), can be fitted by analytical functions. Therefore, the accuracy of the method is only limited by the limit of validity of the linear wave theory. Also the complexity of the calibration increases with the number of parameters A and R to be determined. We will see in the next chapter, that the accuracy and number of experimental data do not justify the use of more than one or two Hankel transforms. The problem then boils down to determining the most appropriate functions which fit most of the data. If only one function is used, then only one value for A and one value for R needs to be determined as a function of the explosion parameters.

The mathematical linear model for EGWWs then exhibits a remarkable simplicity. What is even more remarkable is that such a simple model translates so accurately to such a complex phenomena, particularly in the case of deep and intermediate water depths. In shallow water, the nonlinearity of the problem imposes more severe limitations, as will be seen in Chapter 4.

Table 2.1. Hankel transforms.

Cases	Initial disturbance η_0, w_s, I	Initial conditions η_0, w_s, I	Hankel transform $\bar\eta_0, \bar w_s, I$	Vol.* πR^2	Energy* $2\pi\varrho g R^2$
1	uniform 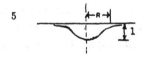	$r_0 \leqq R_0 = R \quad -1$ $r_0 > R_0 = R \quad 0$	$-\dfrac{R}{k} J_1(kR)$	1	$\dfrac{1}{4}$
2	parabolic	$r_0 \leqq R_0 = R \quad \left(\dfrac{r_0}{R}\right)^2 - 1$ $r_0 > R_0 = R \quad 0$	$-\dfrac{2}{k^2} J_2(kR)$	$\dfrac{1}{2}$	$\dfrac{1}{12}$
3	parabolic zero volume	$r_0 \leqq R_0 = R \quad 2\left(\dfrac{r_0}{R}\right)^2 - 1$ $r_0 > R_0 = R \quad 0$	$-\dfrac{R}{k} J_3(kR)$	0	$\dfrac{1}{12}$
4	quadric zero volume	$r_0 \leqq R_0 = \sqrt{3}R$ $-\dfrac{1}{3}\left(\dfrac{r_0}{R}\right)^4 + \dfrac{4}{3}\left(\dfrac{r_0}{R}\right)^2 - 1$ $r_0 > R_0 = \sqrt{3}R \quad 0$	$-\dfrac{4}{k^2} J_4(kR)$	0	$\dfrac{1}{10}$
5	hyperbolic	$-\dfrac{R^3}{(R^2 + r_0^2)^{3/2}}$	$-R^2 e^{-kR}$	2	$\dfrac{1}{8}$

Table 2.1. (*Continued*)

Cases	Initial disturbance η_0, w_s, I	Initial conditions η_0, w_s, I	Hankel transform $\bar{\eta}_0, \bar{w}_s, I$	Vol.* πR^2	Energy* $2\pi \varrho g R^2$
	Gaussian				
6		$-\exp\left[-\left(\dfrac{r_0}{R}\right)^2\right]$	$-\dfrac{R^2}{2}\exp\left[-\left(\dfrac{kR}{2}\right)^2\right]$	1	$\dfrac{1}{8}$
7		$-\dfrac{n!R^{n+1}}{(R^2+r_0^2)^{\frac{n+1}{2}}}$ $\times P_n\left[\dfrac{R}{(R^2+r_0^2)^{1/2}}\right]$	$(kR)^{n-1}\exp(-kR)R^2$		

* These two columns are for free-surface elevation η_0 only.

The general characteristics of EGWWs, such as mathematically formulated herewith, have been presented in Chapter 1. Basically, there are two types of solutions.

The first one is described by a Hankel transform in the form of a Bessel function, in which case the solution appears as a succession of wave trains of decreasing amplitude and duration. (Cases 1–4 of Table 2.1). Figure 1.14 is a typical example of this type of solution, given by a parabolic crater with lip (Case 3).

The second ones are in exponential form (Cases 5–7), for which the wave amplitude decreases monotically. There is no beating and the maximum wave is the leading wave (Figure 1.16).

The first case fits the wave records obtained in deep water rather well. In shallow water, the energy tends to shift towards the leading wave. There are very few waves per wave trains and Cases 5 to 7 are more representative.

Therefore, the second family of mathematical models has been used in the past in the case of explosions in shallow water (Wang *et al.*, 1988). The latest investigations retained here indicate that the best models are uniformly valid for deep and shallow water. They are given by the sum of an upward velocity $w_s(r_0)$ in the form of a dome given by Case 5, and a free-surface elevation

$\eta(r_0)$ in the form of a parabolic crater with lip given by Case 3, or by a quadric (Case 5). In conclusion, the general linear solution for EGWWs is a function of four parameters A_η, R_η, A_w, R_w, such that

$$\eta(r,\,t) = \int_0^\infty \left(A_\eta \frac{R_\eta}{k} J_3(kR_\eta) \cos \sigma t + A_w R_w^2 \exp(-kR_w) \sin \sigma t \right) J_0(kr) k\,dk \,.$$

$$(2.83)$$

Applying the method of the stationary phase, one obtains:

$$\eta(r,\,t) = \frac{1}{r} \left(\frac{kV(k)}{-\dfrac{dV}{dk}} \right)^{1/2} \left[A_\eta \frac{R_\eta}{k} J_3(kR_\eta) \cos(\sigma t - kr) \right.$$

$$\left. + A_w R_w^2 \exp[-kR_w]\sigma \sin(\sigma t - kr) \right] .$$

$$(2.84)$$

The time histories presented in Figure 1.14 are given by this equation where A_w has been set to zero.

8. The Leading Wave

As discussed previously, for shallow water explosions the maximum wave is the first or second wave. The shallow water explosion generated waves generally present no beating after a long distance of propagation. If the investigation is limited to shallow water propagations, an initial exponential source form should be an excellent choice for wave analysis, as it yields a distinctive feature of the leading wave, which dominates the significance over all the trailing waves. In the following, therefore, the theory of the leading wave for an exponential disturbance assimilating an explosion in shallow water is presented.

Consider the most general case (Case 7 of Table 2.1), which defines a surface disturbance of the following form

$$\eta_0(r) = -A \frac{n! R^{n+1}}{[R^2 + r^2]^{\frac{n+1}{2}}} P_n \left[\frac{R}{(R^2 + r^2)^{1/2}} \right] ,$$

$$(2.85)$$

if the maximum displacement at the centerline (GZ) is A. The parameter R represents a characteristic radius of the disturbance and P_n is Legendre's polynomial of degree n.

This initial disturbance has a transform

$$H(k) = -AR^2 (kR)^{n-1} \exp(-kR). \qquad (2.86)$$

Consider the initial disturbance as a static crater, then the resulting wave history is given by Equation (2.31).

$$\eta(r, t) = AR \int_0^\infty (kR)^n \exp(-kR) J_0(kr) \cos \sigma t \, dk. \qquad (2.87)$$

For the leading wave we are looking at mildly dispersive solution for which the dispersive relation can be approximated by the leading two terms of its expansion (2.51),

$$\sigma = k - \frac{k^3}{6}. \qquad (2.88)$$

For the possibility of an analytical approximation, the problem is further limited to a large distance from the source where the Bessel function $J_0(kr)$ can be represented by its asymptotic form (2.54):

$$J_0(kr) = \left(\frac{2}{\pi kr}\right)^{1/2} \cos\left(kr - \frac{\pi}{4}\right). \qquad (2.89)$$

Substituting (2.88) and (2.89) in (2.87), and neglecting the terms of order $(1/r)^{3/2}$ and higher gives

$$\eta(r, t) = A \left(\frac{1}{2\pi r}\right)^{1/2} R^{n+1} \int_0^\infty k^{n-1/2} \exp(-kR) \cos\left(kr - kt + \frac{k^3 t}{6} - \frac{\pi}{4}\right) dk. \qquad (2.90)$$

In complex variable notations, Equation (2.90) may be written as

$$\eta(r, t) = \frac{A}{2} (\pi r)^{1/2} R^{n+1}$$

$$\times Re\left\{ (1-i) \int_0^\infty k^{n-1/2} \exp\left[i\left(kr - kt + \frac{k^3 t}{6}\right) - kR\right] dk \right\}. \qquad (2.91)$$

Using new variables

$$s = \left(\frac{6}{t}\right)^{1/3} (r - t),$$

$$q = \left(\frac{6}{t}\right)^{1/3} R, \qquad (2.92)$$

$$u^2 = \left(\frac{6}{t}\right)^{-1/3} k,$$

Equation (2.91) can be expressed in the following form

$$\eta(r, t) = A\left(\frac{1}{r}\right)^{1/2} \left(\frac{6}{t}\right)^{-1/6} I(s, q, n), \qquad (2.93)$$

where

$$I(s, q, n) = Re\left\{\frac{1-i}{\pi^{1/2}} q^{n+1} \int_0^\infty u^{2n} \exp[i(u^6 + su^2) - qu^2]du\right\}. \qquad (2.94)$$

The integral is affinitive to the Airy integral but no closed form or tabulated solution is yet established. Replacing u by $ve^{i\pi/12}$ in (2.94) yields

$$I(s, q, n) = \frac{2}{\pi} q^{n+1} Re\left\{e^{i(n-1)\pi/6} \int_0^\infty v^{2n} e^{-v^6}\right.$$

$$\left. \times \exp[(s^2 + q^2)^{1/2} e^{i(2\pi/3+\gamma)} v^2]dv\right\}, \qquad (2.95)$$

where $\gamma = \tan^{-1}(q/s)$. By further expanding the exponential function inside the integral into a series, Equation (2.95) may be written as

$$I(s, q, n) = \left(\frac{2}{\pi}\right)^{1/2} q^{n+1} Re\left\{e^{i(n-1)\pi/6} \int_0^\infty e^{-v^6}\right.$$

$$\left. \times \sum_{m=0}^\infty \frac{1}{m!}(s^2 + q^2)^{m/2} v^{2m+2n} e^{im(2\pi/3+\gamma)} dv\right\}. \qquad (2.96)$$

Before integrating the series term by term, setting

$$\eta = v^6, \qquad (2.97)$$

Equation (2.96) yields:

$$I(s, q, n) = \frac{1}{6}\left(\frac{2}{\pi}\right)^{1/2} q^{n+1} Re\left\{ e^{i(n-1)\pi/6} \int_0^\infty e^{-\eta} \right.$$

$$\left. \times \sum_{m=0}^\infty \frac{1}{m!}(s^2+q^2)^{m/2} \eta^{(2m+2n-5)/6} e^{im(2\pi/3+\gamma)} d\eta \right\}. \qquad (2.98)$$

It is known that the gamma function can be expressed by an integral of the following form

$$\Gamma(y) = \int_0^\infty x^{y-1} e^{-x} dx. \qquad (2.99)$$

In terms of the gamma function, the function $I(s, q, n)$ is therefore given by

$$I(s, q, n) = \frac{1}{6}(2\pi)^{1/2} q^{n+1} Re\left\{ e^{i(n-1)\pi/6} \right.$$

$$\left. \times \sum_{m=0}^\infty \frac{1}{m!}(s^2+q^2)^{m/2} e^{im(2\pi/3+\gamma)} \Gamma\left(\frac{2m+2n+1}{6}\right) \right\}. \qquad (2.100)$$

The function $I(s, q, n)$ is unique for a given source function $\eta_0(r)$ and the surface wave motion can be uniquely determined from (2.93). In particular, for $n = 1$ (Case 5)

$$\eta_0(r) = AR^3[R^2+r^2]^{-3/2}, \qquad (2.101)$$

$$\eta(r, t) = A\left(\frac{1}{r}\right)^{1/2}\left(\frac{6}{t}\right)^{-1/6} I(s, q, 1). \qquad (2.102)$$

The waves generated by an initial surface displacement are directly related to the function I as shown in Equation (2.102). Consequently, the behavior of the waves are uniquely determined by the behavior of the function I.

Shown in Figure (2.1) are the variations of I as a function of negative s for various values of q ranging from small to large. The value of negative s corresponds to the time at a given observation point r. The variation along the $-s$ axis for each case therefore may be viewed as a time history of the wave train at a fixed point. The value of $s = 0$ refers to the theoretical arrival time of long waves from the origin. That the leading wave in fact arrives considerably earlier than the theoretical arrival time is partially a result of nonlinear dispersion

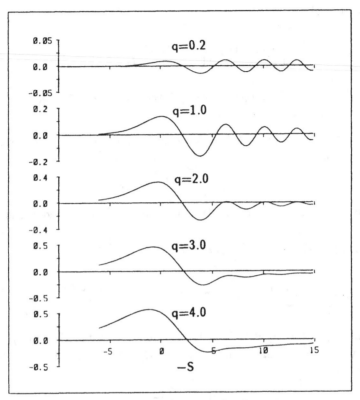

Fig. 2.1. Function I.

effects. The approximation (2.88) for the dispersion relationship is equivalent to adding a nonlinear correction to the long waves which propagate with a velocity equal to the square root of gd.

As shown in these figures, these functions become less and less dispersive when the parameter q increases. This simply indicates that at a given location of observation, waves appear to be more dispersed from small source disturbances and less dispersed from very large source disturbances. On the other hand, if one considered the parameter q as a field parameter for a fixed source radius R, the results clearly demonstrate that waves are nondispersive in the relatively near field but dispersion gradually becomes significant as waves propagate outward. Furthermore, these plots indeed verify that waves are dispersive in deep water (small q) but tend to be less dispersive in very shallow water (large q), if one regards q as a water depth parameter with everything else held fixed.

The same function I may be plotted as a function of positive s as shown in Figure 2.2. Considering the source being located at $s = -\infty$, the plots can be read as snapshots of the waves arriving at different locations along the spacial, radial coordinate, while propagating toward the positive s direction. Viewing from large to small q (considering q as a field parameter), the evolution of waves propagating from near to far from the source is particularly demonstrated by these plots.

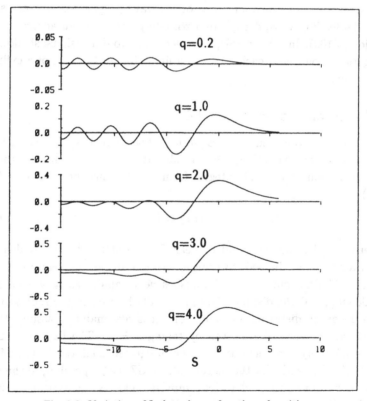

Fig. 2.2. Variation of I plotted as a function of positive s.

It is clear that at the wave front, $r \approx t$. Consequently, Equation (2.102) can be expressed as

$$\eta \sim r^{-1/3} I(s, q, 1) \qquad (2.103)$$

A careful analysis of the function I reveals that the leading wave (first peak) varies as a function of q^2 when q approaches zero. Since q itself varies as $t^{-1/3}$ or $r^{-1/3}$ for a given source dimension R, Equation (2.103) simply indicates that the leading wave varies as r^{-1} when q is small. Therefore, for small source disturbances, for a distance far away from the source, or for propagation in very deep water, the leading wave decays like r^{-1}. However, for relatively large q ($q > 3$), the leading waves behave differently, the function I varies approximately as a linear function of q when q is large. This indicates that for large source disturbances, for a location relatively near to the source region as compared to a distance far away, or for wave propagation in very shallow water, the leading wave decays approximately like $r^{-2/3}$ in accordance with Equation (2.103). In other words, wave decay of two-dimensional shallow water propagation should approach $r^{-2/3}$ as a limit, like in the case of a cylindrical solitary wave.

9. Movement at or Near GZ

The solutions represented by Equation (2.80), which are based on the stationary phase approximation, are not valid at or near GZ. In order to obtain the wave motion at GZ, the integral forms (2.79) must be retained. At GZ, ($r = 0$)

$$J_0(kr) = 1 . \qquad (2.104)$$

Furthermore, the movements at or near GZ are little affected by dispersion, even in deep water. Indeed, to achieve effective dispersion requires time and distance. The movement at GZ vanishes very rapidly as it will be seen in the following, and the distances are practically limited to the radius R of the original localized disturbances. Therefore, it is reasonable to assume that the solutions at or near GZ are given by taking $\sigma \cong k$. The Equation (2.79) is then considerably simplified and can be integrated analytically (Le Méhauté et al., 1987). In particular the movements at GZ, corresponding to the various original conditions (water elevation, impulse, velocity) analytically defined in Table 2.1 (Cases 1–5), are given by

$$\eta(0,\,t) = \int_0^\infty \begin{vmatrix} \dfrac{R}{k} & J_1(kR) \\[2mm] \dfrac{2}{k^2} & J_2(kR) \\[2mm] \dfrac{R}{k} & J_3(kR) \\[2mm] \dfrac{4}{k^2} & J_4(kR) \\[2mm] R^2\,\exp(-kR) \end{vmatrix} \begin{vmatrix} k\,\cos\,kt \\[1mm] \sin\,kt \\[1mm] k^2\,\sin\,kt \end{vmatrix} dk \qquad (2.105)$$

These integrals have been integrated analytically and the results are presented in Table 2.2. In this table, the notation τ is introduced

$$\tau = \frac{t}{r}. \qquad (2.106)$$

The results are presented graphically in Figures 2.3, 2.4, and 2.5 which correspond to the case of an initial crater, initial velocity, and initial impulse respectively. The five curves presented on each figure correspond to the five types of Hankel Transform inserted in Equation (2.105).

It is easily verified that in accordance with Equation (2.35) the curves in Figure 2.5 are the derivative of the curves in Figure 2.3, which themselves are the derivatives of the curves in Figure 2.4.

Fully dispersive solutions (Case 5), namely,

$$\eta(0,\,t) = \int_0^\infty R^2\,\exp(-kR) \begin{vmatrix} \cos\,\sigma t \\[1mm] \dfrac{1}{\sigma}\,\sin\,\sigma t \\[1mm] \sigma\,\sin\,\sigma t \end{vmatrix} k\,dk \qquad (2.107)$$

have also been resolved numerically, thanks to the exponential convergence of the integrand, for a variety of values of the R parameters. The results are shown in Figures 2.6, 2.7, and 2.8 corresponding to the case of initial disturbance in the form of a crater, velocity, and impulse respectively. A comparison between these figures, where $R = 1$ (middle curves), and the curves in Figures 2.3 to 2.5, which correspond to the same Hankel Transform $[R^2\,\exp(-kR)]$ (last curves), substantiates the fact that the dispersive effects are very small at or near GZ.

Table 2.2. Nondispersive solutions at GZ.

Initial conditions				$\tau = \dfrac{t}{R}$	Case 1	Case 2
Free surface	Velocity	Impulse			$-1 \quad r_0 \leq R_0 = R$ $0 \quad r_0 > R_0 = R$	$\left(\dfrac{r_0}{R_0}\right)^2 - 1 \quad r_0 \leq R_0 = R$ $0 \quad r_0 > R$
$\eta(r,0)$	0	0	Integrand		$-RJ_1(kR)\cos kt$	$-2k^{-1}J_2(kR)\cos kt$
			Solution $\eta(0,t)$	$0<\tau<1$	-1	$2(1-\tau^2)-1$
				$1<\tau<\infty$	$\dfrac{-1}{(\tau^2-1)^{1/2}(\tau+(\tau^2-1)^{1/2})}$	$\dfrac{1}{(\tau+(\tau^2-1)^{1/2})^2}$
0	$w_s(r,0)$	0	Integrand		$-Rk^{-1}J_1(kR)\sin kt$	$-2k^{-2}J_2(kR)\sin kt$
			Solution $\eta(0,t)$	$0<\tau<1$	$-R\tau$	$1 - \dfrac{R\tau}{3}[2(1-\tau^2)+1]$
				$1<\tau<\infty$	$\dfrac{-R}{\tau+(\tau^2-1)^{1/2}}$	$-\dfrac{R(\tau+2(\tau^2-1)^{1/2})}{3(\tau+(\tau^2-1)^{1/2})^2}$
0	0	$I(r,0)$	Integrand		$-RkJ_1(kR)\sin kt$	$2J_2(kR)\sin kt$
			Solution $\eta(0,t)$	$0<\tau<1$	0	0
				$1<\tau<\infty$	$\dfrac{[2\tau+(\tau^2-1)^{1/2}]}{R(\tau^2-1)^{3/2}[\tau+(\tau^2-1)^{1/2}]}$	$\dfrac{[2\tau+(\tau^2-1)^{1/2}]}{R(\tau^2-1)^{1/2}[\tau+(\tau^2-1)^{1/2}]^2} - \dfrac{4\tau}{R}$

Table 2.2. (*Continued*)

Case 3	Case 4	Case 5
$2\left(\dfrac{r_0}{R_0}\right)^2 - 1 \quad r_0 \leqq R_0 = R$ $0 \qquad r_0 > R_0 = R$	$\dfrac{1}{3}\left(\dfrac{r_0}{R_0}\right)^4 + \dfrac{4}{3}\left(\dfrac{r_0}{R_0}\right)^2 - 1 \quad r_0 \leqq \sqrt{3R_0} = R$ $0 \qquad r_0 \leqq \sqrt{3R_0} = R$	$\dfrac{-R^3}{(R^2+r_0^2)^{3/2}} \quad R_0 = R$
$-RJ_3(kR)\cos kt$	$-4k^{-1}J_4(kR)\cos kt$	$-R^2 k e^{-kR}\cos kt$
$0 < \tau < 1 \quad -4(1-\tau^2)+3$ $1 < \tau < \infty \quad \dfrac{-1}{(\tau^2-1)^{1/2}(\tau+(\tau^2-1)^{1/2})^3}$	$0 < \tau < 1 \quad -8(1-\tau^2)^2 + 8(1-\tau^2) - 1$ $1 < \tau < \infty \quad \dfrac{-1}{(\tau+(\tau^2-1)^{1/2})^4}$	$\dfrac{1-\tau^2}{(1+\tau^2)^2}$
$-Rk^{-1}J_3(kR)\sin kt$	$-4k^{-2}J_4(kR)\sin kt$	$-R^2 e^{-kR}\sin kt$
$0 < \tau < 1 \quad -R\tau + \dfrac{4}{3}R\tau$ $1 < \tau < \infty \quad \dfrac{R}{3(\tau+(\tau^2-1)^{1/2})^3}$	$0 < \tau < 1 \quad -\left[\dfrac{4R}{15}(1-\tau^2)^{1/2}\sin(4\sin^{-1}\tau)\right.$ $\left.\qquad\qquad - \dfrac{R\tau}{15}\cos(4\sin^{-1}\tau)\right]$ $1 < \tau < \infty \quad \dfrac{R[\tau+4(\tau^2-1)^{1/2}]}{15[\tau+(\tau^2-1)^{1/2}]^4}$	$-\dfrac{R\tau}{1+\tau^2}$
$-RkJ_3(kR)\sin kt$	$-4J_4(kR)\sin kt$	$-R^2 k^2 e^{-kR}\sin kt$
$0 < \tau < 1 \quad \dfrac{8\tau}{R}$ $1 < \tau < \infty \quad \dfrac{-2\tau(\tau^2-1)^{1/2}}{R(\tau^2-1)^{3/2}[\tau+(\tau^2-1)^{1/2}]}$	$0 < \tau < 1 \quad -\dfrac{4\sin(4\sin^{-1}\tau)}{R(1-\tau^2)^{1/2}}$ $1 < \tau < \infty \quad \dfrac{4}{R(\tau^2-1)^{1/2}(\tau+(\tau^2-1)^{1/2})^4}$	$-\dfrac{4R^{1/2}(3-\tau^3)}{(1+\tau^2)^3}$

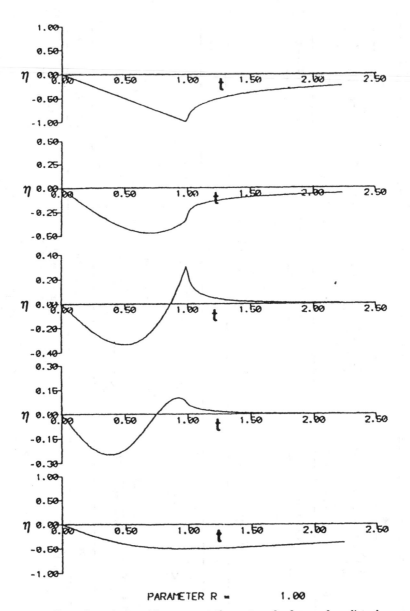

PARAMETER R =   1.00

Fig. 2.3. Nondispersive solution. Movement at the center of a free-surface disturbance as a function of the dimensionless time t, defined by an *initial crater* corresponding to Cases 1–5 in Table 2.1, $R = 1$.

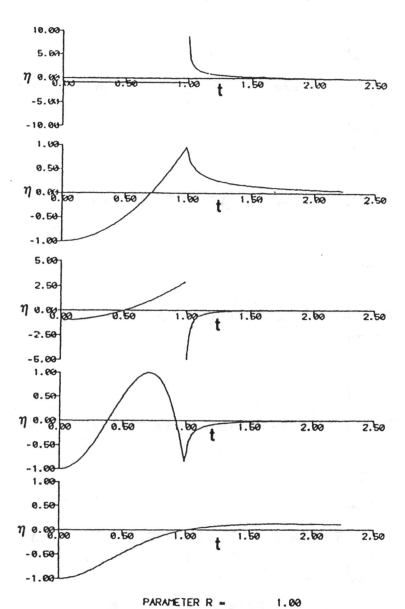

PARAMETER R = 1.00

Fig. 2.4. Nondispersive solution. Movement at the center of a free-surface disturbance as a function of the dimensionless time t, disturbance defined by an *initial downward free-surface velocity* corresponding to Cases 1–5 in Table 2.1, $R = 1$.

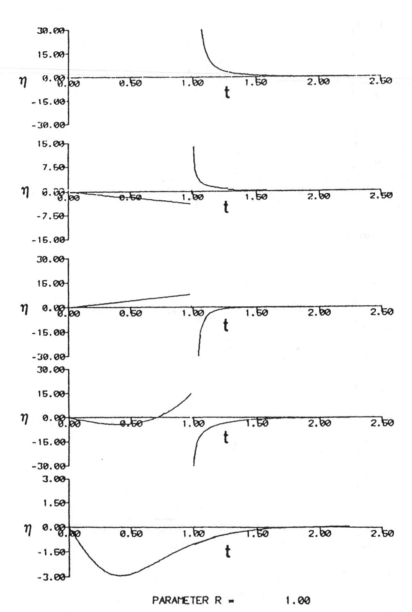

PARAMETER R =   1.00

Fig. 2.5. Nondispersive solution. Movement at the center of a free-surface disturbance as a function of the dimensionless time t, defined by an *initial impulse* on the free surface corresponding to Cases 1–5 in Table 2.1, $R = 1$.

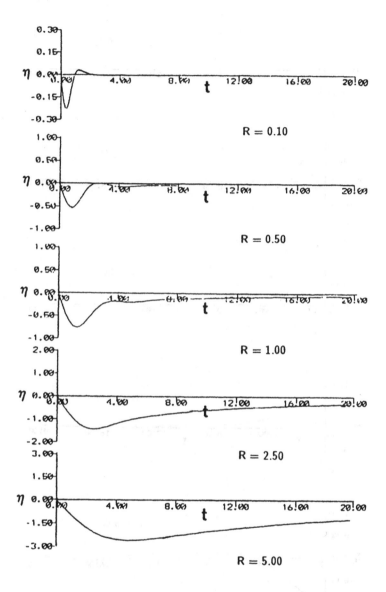

DISTANCE FROM G.Z. = 0.00

Fig. 2.6. Dispersive solution. Movement at the center of a free-surface disturbance as a function of the dimensionless time t, defined by an *initial crater* in Case 5 in Table 2.1, for various values of R.

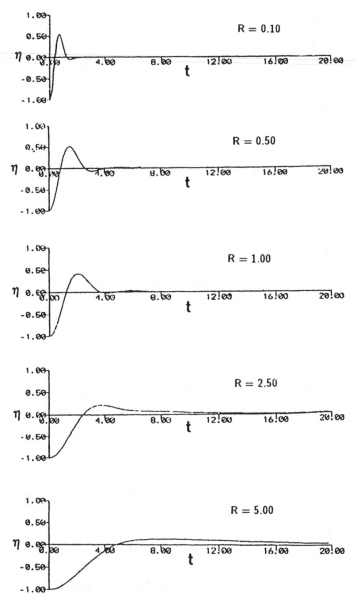

Fig. 2.7. Dispersive solution. Movement at the center of a free-surface disturbance as a function of the dimensionless time t, defined by an *initial velocity* in Case 5 in Table 2.1, for various values of R.

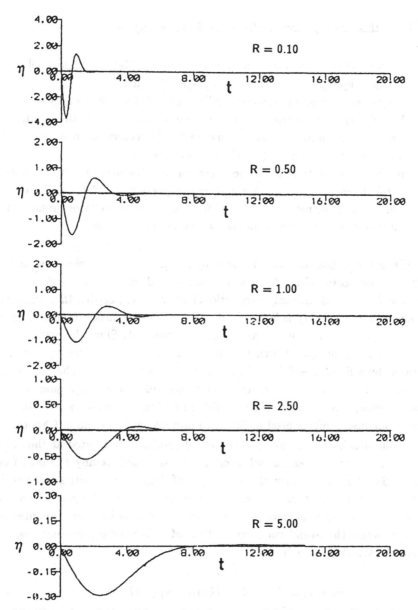

Fig. 2.8. Dispersive solution. Movement at the center of the free-surface disturbance as a function of the dimensionless time t, defined by an *initial impulse* in Case 5 in Table 2.1, for various values of R.

From this investigation the following features appear:

(1) It is seen that all the solutions are nearly nonoscillatory as most solutions yield a unique peak-trough (or trough-peak) solution.

(2) In many cases where the initial conditions exhibit a discontinuity, it is seen that $\eta(0, t) \to \infty$ when $\tau \to 1$. In Case 5 where $n = 1$, the continuous nature of the initial condition and of its derivative ensures continuous solutions for $\eta(r, t)$ and that $\eta(0, t)$ is always finite.

(3) In shallow water, the free-surface elevation $\eta(r, t)$ due to an initial elevation $\eta_0(r_0)$ depends solely upon τ and is independent of R for a constant τ value. When the movement results from an initial velocity or pressure, the solution $\eta(0, t)$ is function of the two parameters R and τ.

When the initial condition is given by a velocity w_s, the elevation at GZ, $\eta(0, t)$, is proportional to R (for a constant τ) and when they are given by an impulse I, $\eta(0,t)$ is inversely proportional to R. It is possible that when the extent of initial velocity is large (large R) or when the initial impulse is small (small R), $\eta(0, t)$ can theoretically be smaller than -1. Since the seafloor is at $z = -1$, $w_{s\ max}$ or I_{max}, which here has been taken equal to unity, should be reduced by a linear coefficient A so that this does not occur. The case where $\eta_{max}(0, t)$ is less than -1 is nevertheless possible for an explosion in very shallow water, inducing ground cratering in addition to water waves. In such a case, nonlinear effects need to be taken into account as done in Chapter 4.

In the previous analysis, the theoretical results were limited at the origin ($r = 0$). Since the general solutions (2.79) are valid at any distance from the origin, the time history of the collapse of domes and cavities can also be obtained. This can also be done either numerically in the general case or analytically in the nondispersive case. Analytical solutions are presented for Case 5, when the initial condition is that of a free-surface elevation $\eta_0(r_0)$. Then the solution for $\eta_\eta(r, t)$ is

$$\eta_\eta(r, t) = \int_0^\infty k\, dk\, J_0(kr) R^2 \exp(-kR) \cos \sigma t\,, \qquad (2.108)$$

which can be integrated analytically in the nondispersive case where $\sigma = k$, to give:

$$\eta(r,\,t) = \frac{R^3}{\sqrt{2}}\frac{2ac + cd - d^2}{d^3(a+d)^{1/2}}\,, \qquad (2.109)$$

where

$$a = R^2 + r^2 - t^2$$

$$b = 2Rt$$

$$\qquad\qquad (2.110)$$

$$c = R^2 + r^2 + t^2$$

$$d = (a^2 + b^2)^{1/2}\,.$$

It is verified that when $r = 0$, Equations (2.109) and (2.110) yield the same solution as the one presented in Table 2.2, Case 5, $\eta_0(r_0)$.

The corresponding time history of the free-surface elevation is presented in Figure 2.9 in the cases where $R = 0.1$, 1, and 5, where the nondimensional time step is 0.5. The nonoscillatory nature of the movement is confirmed. The same equation (2.107) has also been integrated numerically in the dispersive case ($\sigma \neq k$). The lack of dispersion near the origin is also confirmed.

In summary, the investigation of the wave motion at or near GZ shows the following:

(1) It confirms that the impulse solution is the time derivative of the elevation solution which is the time derivative of the velocity solution.
(2) The solutions are either nonoscillatory or quasi-nonoscillatory at the origin.
(3) The solution near the origin are little affected by frequency dispersion even in deep water.
(4) The multiplicity of radiating, circular waves observed at a distance is solely the result of frequency dispersion from that unique oscillation at the origin.
(5) The solutions, even though linear, give apparently realistic results even in the cases where the movement should, *a priori*, be highly nonlinear.
(6) The vertical projection of water at the origin in the form of a jet, a spike, or a drop is explained theoretically by discontinuity in the initial conditions.
(7) The qualitative comparisons between theoretical results and examples of physical phenomena lead to the conclusion that linear calibration of the initial disturbance is possible in specific cases where a quantitative prediction is needed.

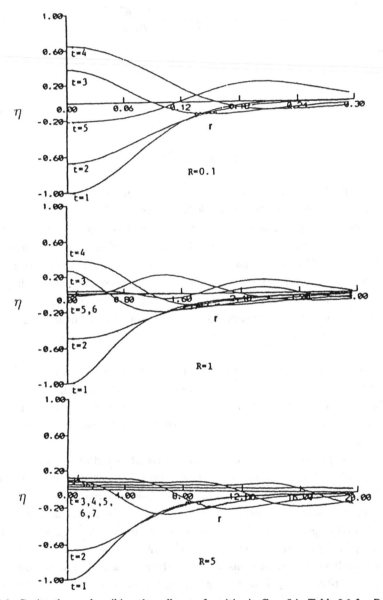

Fig. 2.9. Cavity shapes describing the collapse of cavities in Case 5 in Table 2.1 for $R = 0.1$, 1, and 5.

10. Limits of Validity

The approximations made in various versions of solutions involve essentially two assumptions on the relevant variables: the assumption of a relatively simple dispersion relationship and the assumption of a distance far from the source region. In addition, the variable of time is implicitly involved through the variable of distance. In order to identify the various regions of validity for the various approximation methods, three characteristic parameters are defined.

(1) *The dispersion (or water depth) parameter*

The dispersion characteristics are governed by a single variable, the wave number k. Since the wave is directly related to the radius of the disturbance, a dispersion parameter δ is defined as the ratio of the water depth to the characteristic radius:

$$\delta = \frac{1}{R}. \tag{2.111}$$

It is evident that the variation of this parameter from small to large would indicate the significance of the dispersion or the water depth effect. One should also realize that the inverse of this parameter, $1/\delta$, may be viewed as a source dimension parameter when such an interpretation is needed.

(2) *The distance parameter*

The argument kr governs the applicability of the asymptotic form of the Bessel function implemented in the mildly dispersive and stationary phase solutions. Again relating the wave length to the source disturbance radius, a distance parameter λ is defined as the ratio between the distance and the characteristic radius,

$$\lambda = \frac{r}{R}. \tag{2.112}$$

This indicates that the relative length of the stand-off distance with respect to the source of disturbance dimension in fact dictates the validity of the distance approximation.

(3) *The time parameter*

The time of arrival of an individual portion of a wave system constitutes in part the characteristics of propagation of that particular distance. Consequently, a time parameter τ is defined as the ratio between the elapsed time and the stand-off distance,

$$\tau = \frac{t}{r}. \tag{2.113}$$

It is clear that $\tau = 1$ corresponds to the theoretical long wave arrival time from $r = 0$.

It has been shown that the nondispersive solution is valid at a close distance from the origin when the disturbance is large. In order to get a quantitive statement of how close the distance and how large the source radius should be, one must examine a great number of distances close to the origin and a great number of source radii, Rs. A better way to identify these ranges of validity is by using the parameter q which has been established in the mildly dispersive solution.

The parameter q can be approximately written as

$$q \simeq \left(\frac{6}{r}\right)^{1/3} R, \tag{2.114}$$

at the wave front where $r \simeq t$. Consequently, this parameter would provide a guide to the distance and source radius. Further, referring to Equation (2.92), the negative s may be viewed as a time variable at a fixed observation point.

The nondispersive solution, the numerically obtained fully dispersive solution, and the mildly dispersive solution are plotted on a common basis of $-s$ for several values of q and compared in Figure 2.10.

By comparing the nondispersive results with the fully dispersive solution, one may find that the nondispersive solution is sufficiently accurate when q is larger than 4, or

$$r < \frac{3}{32R^3}. \tag{2.115}$$

In terms of parameters δ and λ, the relation for the nondispersive solution to be valid may be written as

$$\delta^2 < \frac{3}{32}. \tag{2.116}$$

The time parameters seems to be irrelevant in the nondispersive case when q is large.

The validity in the prediction of the leading wave by the mildly dispersive solution is confirmed again by comparing it with the fully dispersive solution in Figure 2.10. The agreement for all values of q between the two solutions indicates again that the mild dispersion assumption (2.88) is valid for leading wave approximations. The only constraint to the leading wave solution is,

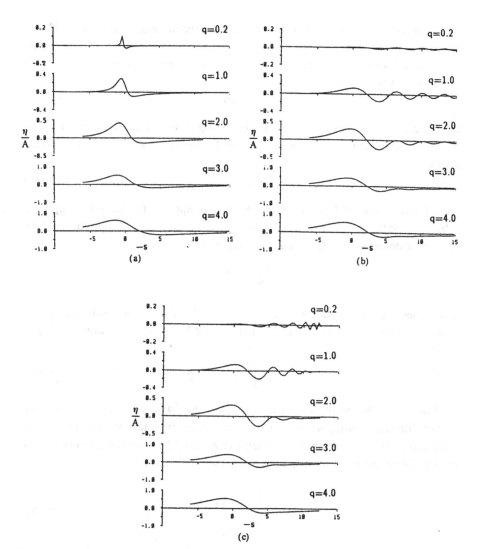

Fig. 2.10. Comparison of (a) nondispersive, (b) mildly dispersive, and (c) fully dispersive solutions in terms of parameters s and q.

therefore, the distance requirement imposed by the approximation of the Bessel function. It is known that the zero order Bessel function can be approximated by its asymptotic expression with sufficient accuracy if its argument is greater than 10. In the present case,

$$kr > 10 . \tag{2.117}$$

On the other hand, the period T of the leading wave is known to vary proportionally to the cube root of the distance and is approximately given by (Wang, 1987)

$$T \simeq (6^2 r)^{1/3} . \tag{2.118}$$

Also, for the range of k where the leading wave applies, the long wave approximation is valid and the dimensionless wave length can be approximated by the dimensionless period, so that one may write

$$k \simeq \frac{2\pi}{T} . \tag{2.119}$$

Consequently, the constraint imposed to the mildly dispersive solution can be derived by inserting Equations (2.118) and (2.119) into Equation (2.117) to yield

$$r > 6\left(\frac{10}{2\pi}\right)^{3/2} \simeq 12 . \tag{2.120}$$

Therefore, the mildly dispersive solution should provide an accurate prediction for the leading wave at a distance of approximately 12 depths away from the active source region. In terms of δ, λ, and τ, the constraints for the mildly dispersive solution can be written as

$$\frac{\lambda}{\delta} > 12 ,$$
$$\tau \simeq 1 . \tag{2.121}$$

As for the method of stationary phase, the constraint of Equation (2.118) imposed by the approximation of the Bessel function is equally valid. As discussed earlier, this method is valid for waves trailing behind the leading wave and should provide a good approximation for motions in relatively deep water. On the other hand, this method provides good predictions for the trailing waves in shallower water as well. However, this is not really interesting because in shallow water propagation, the energy content of all waves longer than the water depth is confined in the leading wave and little is left in the trailing waves. From a mathematical point of view, the method is valid nonetheless, as long as the interest is not focused on the leading wave. By comparing the stationary phase solution with the numerically obtained fully dispersive solution for a series of cases through varying the radius R between 1 and 5, the following empirical criteria are established for the stationary phase method to be valid:

$$(\lambda + 12.5)\delta > 22.5 \,,$$

$$\tau > 1 \,. \tag{2.122}$$

This relation indicates that the dispersion parameter δ can be balanced by the distance parameter λ, and the stationary phase solution would provide a good prediction of the trailing waves even for shallower water large disturbances, but at a far distance.

Chapter 3

CALIBRATION OF THE LINEAR WAVE THEORY

1. Review of Past Investigations — Empirical Correlations

The purpose of this chapter is to calibrate the linear wave theory, presented in Chapter 2, by means of comparing the theoretical time-history of the free-surface elevation for a fictitious initial disturbance with the experimental results generated by factual explosions.

For deep-water explosions, experimental data obtained through several series of explosion tests using chemical charges at the Waterways Experimental Station (Pinkston, 1964; Pace *et al.*, 1970) and Mono Lake (Pinkston and Skinner, 1970) provided the following empirical relations for the maximum amplitude $\eta_{\max}$ and the corresponding period T as summarized by Le Méhauté (1970):

$$\left. \begin{array}{l} \eta_{\max} r = 18\ W^{0.54} \\[2mm] T = 1.669\ W^{0.15} \end{array} \right\} \quad 0 < \frac{z}{W^{0.3}} < 0.25 , \qquad (3.1a)$$

$$\left. \begin{array}{l} \eta_{\max} r = 10\ W^{0.54} \\[2mm] T = 1.773\ W^{0.15} \end{array} \right\} \quad 0.25 < \frac{z}{W^{0.3}} < 7.5 , \qquad (3.1b)$$

111

where W is the yield of the explosive in pounds of TNT and z is the charge position relative to the free surface in feet, positive downward. In this correlation the numerical coefficients are not dimensionless. The correlation relationship (3.1a) applies to bursts which are very close to the free surface, commonly referred to as the upper critical depth, as described in Chapter 1, due to the fact that the wave generation efficiency is highest in this region. Smaller waves would result from bursts at a deeper submergence. However, test results exhibited a second hump that occurred in the middle of the burst region as specified by Equation (3.1b), which is known as the lower critical depth. Insufficient data exist for bursts at depths lower than $z = 7.5\ W^{0.3}$. The product of $\eta_{\max} r$ obtained from the original data is exhibited in Figure 1.8. Using Equations (3.1a) and (3.1b), a correlation of A (crater height) and R (crater radius) is given by Le Méhauté (1970):

$$\left.\begin{aligned} A &= 3.07\ W^{0.24} \\ R &= 9.54\ W^{0.3} \end{aligned}\right\} \quad 0 < \frac{z}{W^{0.3}} < 0.25\,, \tag{3.2a}$$

$$\left.\begin{aligned} A &= 1.51\ W^{0.24} \\ R &= 10.77\ W^{0.3} \end{aligned}\right\} \quad 0.25 < \frac{z}{W^{0.3}} < 7.5\,, \tag{3.2b}$$

These relations suggest that parameters A and R are proportional to $W^{0.24}$ and $W^{0.30}$ respectively. These correlations are purely empirical and no interpretation based on consistent dimensional analysis was pursued in the original research.

The data available for shallow water explosions are very limited. The small charge tests conducted by the Waterways Experiment Station (1955 and 1986) are probably the only source providing wave data with systematic variations of charge weight and water depth. Other information include two high-explosive events fired at the bottom in shallow water at Mono Lake (Whalin, 1966) and a 23-KT mid-depth nuclear burst of Baker at Bikini (DNA EM1, 1978). In addition, the Navy's Project Heat (Willey and Phillips, 1968) provided some information of the plume diameter but no wave measurements were made. In the following discussion, only data obtained from the tests of chemical explosives are considered; nuclear data are not included.

By means of small scale charges (0.5–2048 pounds), the 1955 WES program was designed to estimate wave effects from a 20-KT explosion in water 30–200 feet deep. The charge position varied from beneath the bottom to above the free surface. The results show that the variation of charge location z to water depth ratio from 0 to 1.0 has no well-defined effect on wave height, except in the case of near-surface bursts, where wave generation decreases significantly. The 1986 WES program included tests of explosions with yield ranging from 10–50 pounds in 2.7 feet and 1 foot of water with the purpose of obtaining the complete time history of the free surface at various distances from GZ. Again, no consistent dependence on the depth of burst was observed.

One significant difference in shallow water wave propagation is the dispersion law. As seen in Chapter 2, recall that in deep water, wave height varies inversely with radial distance r as a combined result of frequency and radial dispersions. In extremely shallow water, however, one has seen that the large leading wave is expected to behave like a nondispersive solitary wave as a limit and its height is to vary inversely as $r^{2/3}$ instead of r. It follows that the following decay law should seemingly hold

$$\eta r^{\alpha} = \text{constant} \qquad \frac{2}{3} \leq \alpha(d,\, W) \leq 1. \tag{3.3}$$

Here, α is a function of the water depth d and yield W; the decay ratio α is to vary between 2/3 and 1 should the extrapolation be extended to deep water.

An attempt by Wang *et al.* (1977) to correlate the WES 1955 test data with the largest high-explosive events of Mono Lake gives an empirical relationship

$$\frac{\eta_{\max} r^{\alpha}}{W^{\beta}} = C_W, \tag{3.4}$$

where $\eta_{\max}$ represents the maximum wave amplitude (feet) in the wave train generated by an explosion of yield W (pounds TNT) passing at a distance of r feet. The coefficients α, β, and C_W are given by

$$\alpha = 0.83 \left(\frac{d}{W^{1/3}} \right)^{0.07}$$

$$\beta = \frac{\alpha}{3} + 0.25 \tag{3.5}$$

$$C_W = 1.44 \left(\frac{d}{W^{1/3}} \right)^{0.93},$$

where d is water depth in feet. The above relations show that all these coefficients can be expressed as a function of the water depth to charge yield ratio, $d/W^{1/3}$.

Several important features are presented in this empirical formula. First, the empirical formula embraces both dynamic and energy scalings. It is well known that the energy released by a conventional explosion is proportional to its volume, or the cube of a linear scale. From dimensional analysis, it may be expressed by the following dimensionless parameter, as seen in Chapter 1 (1.16):

$$\pi_1 = \frac{pL^3}{E}, \tag{3.6}$$

It is recalled that p denotes the overburden pressure exerted on the crater boundary or the cavity-fluid interface and L represents a characteristic length. Here, E is the yield in terms of energy released.

On the other hand, one has also seen in Chapter 1 that the total energy of the surface wave required for geometrical similarity must be proportional to $\eta_0^2 R^2$, the potential energy of the crater, and must follow a quadruple scaling which may be expressed by another dimensionless parameter (1.17):

$$\pi_2 = \frac{\rho g L^4}{E}, \tag{3.7}$$

in which ρ is the fluid density and g the gravitational constant.

Since W has been defined in pounds of TNT, one may write

$$E = \mu W, \tag{3.8}$$

where μ is an energy-weight conversion factor of dimension L and may be interpreted as a specific energy. The two parameters π_1 and π_2 then can be written as

$$\pi_1 = \gamma_1 \frac{L^3}{W} \tag{3.9}$$

$$\pi_2 = \gamma_2 \frac{L^4}{W}, \tag{3.10}$$

where γ_1 and γ_2 are two physical parameters of the fluid and the explosives defined as

$$\gamma_1 = \frac{p}{\mu} \tag{3.11}$$

$$\gamma_2 = \frac{\rho g}{\mu}. \tag{3.12}$$

If the effect of charge position in shallow water could be ignored as suggested by the WES 1955 and 1986 data, the pressure p should be approximately constant. Under this postulation, if the conversion factor μ is constant for a given explosive, the parameters γ_1 and γ_2 must be invariant. Thus, for a given yield W, the linear dimensions of the cavity should be proportional to $W^{1/3}$, but the wave height must vary like $W^{1/4}$. The empirical formula indeed supports this fact, and it can be demonstrated more clearly if one rewrites the empirical formula (3.5) in the following form:

$$\bar{\eta}\bar{r}^\alpha = C_W ,\qquad (3.13)$$

where

$$\bar{\eta} = \frac{\eta_{\max}}{W^{1/4}}$$

$$\bar{r} = \frac{r}{W^{1/3}} . \qquad (3.14)$$

Subject to γ_1 and γ_2 being constant, $\bar{\eta}$ varies inversely as $\bar{r}^\alpha$ as suggested by the law of decay postulated in Equation (3.3). In particular, the decay constant α approaches 2/3 as a limit when $d/W^{1/3}$ becomes small. This may also explain why the commonly used formulae in Equations (3.1a) and (3.1b) do not always give satisfactory predictions even for relatively deep water explosions because it fails to take into account the most important parameters for scaling, the depth-yield ratio, which in turn governs the other important feature, the wave decay.

The 1955 WES data cover only the range $0.088 < d/W^{1/3} < 0.585$. The 1986 and 1988 data extended the coverage to $d/W^{1/3} = 1.89$. Assuming that the results are extrapolatable, the following limits for the parameters α, β, and C_W in the empirical formula (3.5) may be obtained

$$\left.\begin{array}{l} \dfrac{2}{3} \leq \alpha \leq 1 \\[2mm] 0.47 \leq \beta \leq 0.58 \\[2mm] 0.079 \leq C_W \leq 17.12 \end{array}\right\} \quad \text{for} \quad 0.044 \leq \frac{d}{W^{1/3}} \leq 14.32 . \qquad (3.15)$$

In particular, at the deep water limit, Equation (3.5) has the following extrapolated result:

$$\frac{\eta_{\max} r}{W^{0.58}} = 17.12 , \qquad (3.16)$$

which is very similar to the formula for upper critical depth established long ago for deep water scaling (3.1a), i.e.,

$$\frac{\eta_{max}T}{W^{0.54}} = 18.$$ (3.17)

2. Generalized Model Based on Inverse Correlation

One has seen in Chapter 2 that, given an experimental wave record $\eta(t)$ at a distance r from GZ, it is possible to determine the initial fictitious linear disturbance which would simulate a similar time history (see formulae 2.32 to 2.78). The method was used based on a series of explosion tests of small charges in shallow to intermediate depths of water conducted at WES (1988 and 1989). To demonstrate the inverse identification process, a particular shot of 10 pounds charge in 4 feet of water at 0.6 feet below the surface is used for a more detailed discussion.

The wave data of this particular shot at a fixed distance $r = 50.2$ feet from GZ is shown in Figure 3.1. The direct Fourier transform of this data gives the real and imaginary components $F_1(\sigma)$ and $F_2(\sigma)$ respectively, according to Equations (2.49) and (2.50), and (2.72) and (2.73). In Figure 3.2, (a) presents the product of the real component $F_1(\sigma)$ and the group velocity V as a function of frequency σ. Similarly, (b) presents the product $V\sigma F_2(\sigma)$. The former

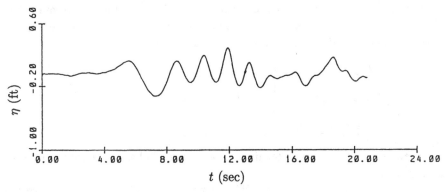

Fig. 3.1. Wave data of 10-pound explosion in 4 feet of water recorded at a distance of 50.2 feet, and the depth of charge = 0.6 feet.

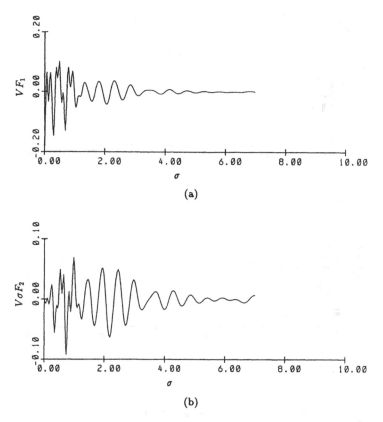

Fig. 3.2. Real and imaginary components of the Fourier transform.

corresponds to the product of $\pi k H_\eta(k) J_0(kr)$ and the latter $\pi k H_w(k) J_0(kr)$, as seen in Equations (2.74) and (2.76). The high frequency oscillations should correspond to the Bessel function $J_0(kr)$ if the linear theory is exactly valid. Comparing the zero crossings of Figure 3.2 with that of $J_0(kr)$ substantially confirms the above statement, although not exactly. Ignoring the irregularities appearing in the oscillations, which are probably due to, to some extent, nonlinearity, the envelope of the oscillations should approximately correspond to $(2\pi k/r)^{1/2} H_\eta(k)$ and $(2\pi k/r)^{1/2} H_w(k)$ if r is sufficiently large.

Figures 3.3 and 3.4 show the two corresponding products multiplied by $r^{1/2}$, derived from actual wave data collected at $r = 32.7'$, $45.2'$, and $50.2'$ from the origin. The similar envelopes shown by the results indeed support

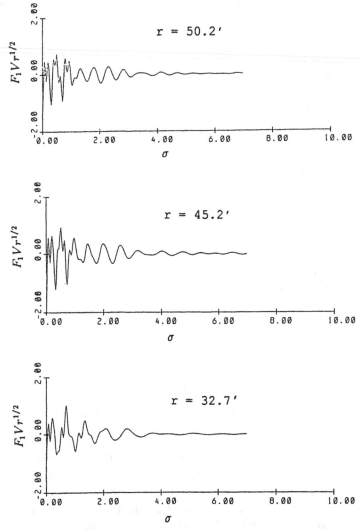

Fig. 3.3. Real parts of the transform of records collected at three different locations.

that $H_\eta(k)$ and $H_w(k)$ are unique, regardless of the location where the input data are collected.

Integrating $H_\eta(k)$ and $H_w(k)$ according to Equations (2.75) and (2.78) yields the initial disturbances $\eta_0(r)$ and $w_s(r)$ respectively. The numerically

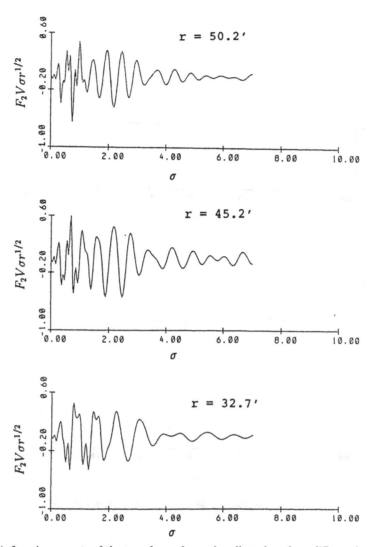

Fig. 3.4. Imaginary parts of the transform of records collected at three different locations.

integrated results using the envelopes presented in Figure 3.2 are shown in Figure 3.5.

A direct forward procedure using $\eta_0(r)$ and $w_s(r)$ obtained above yields a time history of wave elevation $\eta(r, t)$ at any distance r. In order to demonstrate

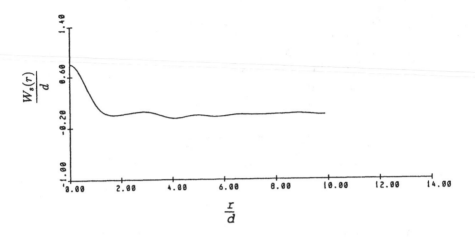

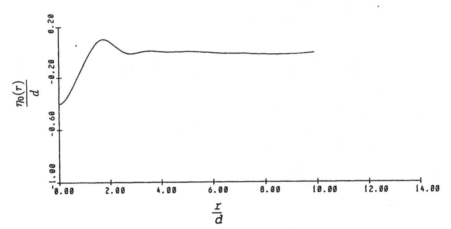

Fig. 3.5. Initial velocity distribution and the crater shape obtained by the inverse method.

the validity of this process, the computed results for $r = 50.2$ feet are compared with the actual record and presented in Figure 3.6. Shown in Figure 3.7 are the components due to $\eta_0(r)$ and $w_s(r)$, obtained by integrating each of them separately. Using the same disturbance functions, $\eta_0(r)$ and $w_s(r)$, waves at other distances are computed and compared to actual measurements (Figure 3.8).

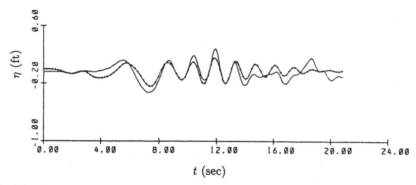

Fig. 3.6. Comparison between the computed result (dotted curve) with the actual wave records for $r = 50.2$ feet.

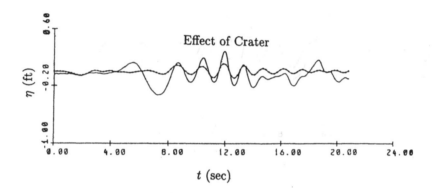

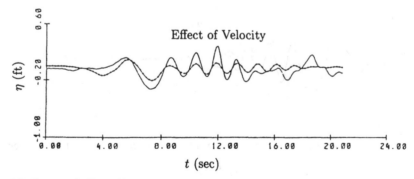

Fig. 3.7. Separated effect of initial conditions $\eta_0(r)$ and $w_s(r)$ on wave form as compared to actual wave data.

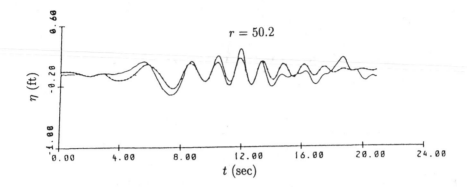

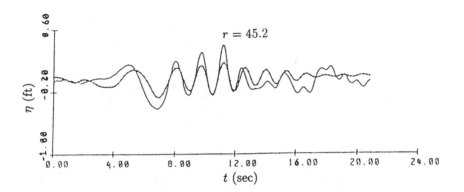

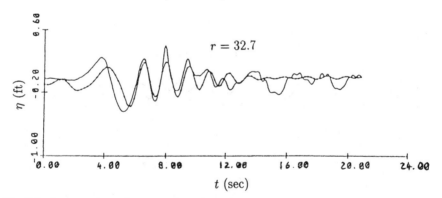

Fig. 3.8. A comparison of computed results with measured data at various locations using initial conditions inversely transformed from data of $r = 50.2$ feet.

Similar results at different spatial locations substantiates the validity of the process.

All the results have been presented by Wang *et al.* (1989) and Khangaonkar and Le Méhauté (1991). As discussed in those papers, the success of this method depends on the accuracy in determining the source disturbance, or more precisely, the envelope functions of $F_1(\sigma)$ and $F_2(\sigma)$ in the inverse process. Direct forward integration of these functions would yield the time history $\eta(r, t)$ exactly, although the source functions $\eta_0(r)$ and $w_s(r)$ obtained through this process represent only a mathematical equivalent, not the actual disturbance.

The source disturbance functions $\eta_0(r)$ and $w_s(r)$ or their transforms $H_\eta(k)$ and $H_w(k)$ derived from the envelope functions of $F_1(\sigma)$ and $F_2(\sigma)$ are a function of yield, water depth, and burst depth in the physical process. In general, if there is sufficient data on $\eta(r, t)$, the geometric characteristics of $\eta_0(r)$ and $w_s(r)$ can be correlated with those physical parameters through the inverse process. For instance, the source disturbances for the case of a ten-pound explosion in 4 feet of water presented in Figure 3.5 can be mathematically approximated by a combination of two mathematical expressions:

$$\eta_0(r) = \begin{cases} A_\eta \left[2\left(\dfrac{r}{R_\eta}\right)^2 - 1 \right] & r \le R_\eta \\ 0 & r > R_\eta \end{cases}$$

(3.18)

$$w_s(r) = \dfrac{A_w}{\left[1 + \left(\dfrac{r}{R_w}\right)^2 \right]^{3/2}},$$

and their corresponding transforms are

$$H_\eta(k) = -A_\eta \dfrac{R_\eta}{k} J_3(kR_\eta)$$

(3.19)

$$H_w(k) = A_w R_w^2 e^{-kR_w}.$$

By fitting the results obtained from the inverse process as shown in Figure 3.9, the four parameters A_η, R_η, A_w and R_w are determined as follows:

$$A_\eta = 0.403, \qquad R_\eta = 1.767,$$

$$A_w = 0.803, \qquad R_w = 0.8.$$

A generalization of the mathematical model would require correlation of the four parameters as a function of the physical parameters of explosions, primarily the yield, water depth, and burst depth. The difficulty in achieving this goal at the present moment is partly due to numerical difficulty in obtaining a unique and accurate solution of the inverse problem and partly because there is no sufficient data to support the analysis.

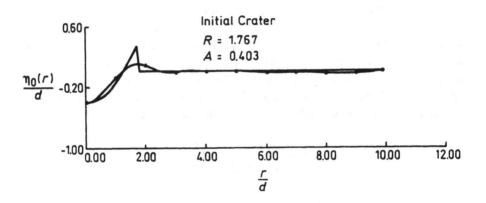

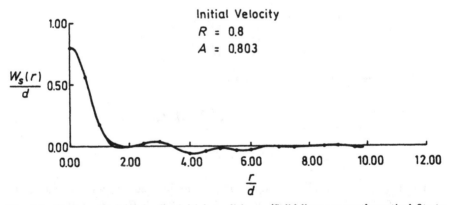

Fig. 3.9. Mathematical fits for the initial conditions. (Solid lines are mathematical fits to the dotted lines obtained by inverse method.)

3. Simplified Model

While the inverse method is potentially powerful in data analysis for many problems, it is unlikely to be a satisfactory tool for phenomenon correlation or scaling of the explosion-generated water waves until uncertainties involved in the physical process can be resolved. The method relies strictly on the information provided by experimental data. Thirty to forty percent or even 100% discrepancy in wave measurements from experiments of supposedly identical explosions is not unusual. Consequently, any effort trying to improve the numerical accuracy involved in the correlation process beyond the limit of experimental uncertainty would be fruitless and even meaningless.

By the inverse method outlined above, the correlation process requires four parameters to describe the equivalent source disturbance, two for the real part of the equation and two for the imaginary part. The former two describe an equivalent water surface depression and the latter two describe the initial velocity of the surface movement. Balancing the error involved from possible numerical inaccuracy in four parameter correlations versus the possible uncertainty involved in the physical process, a two-parameter mathematical model is believed to be sufficient under the present state of the art in describing the actual phenomenon.

Past experience indicates that a source model with initial surface depressions has been satisfactory for correlating waves to deep or shallow water explosions at a far distance. To formulate a uniformly valid model, initial surface depression of the type (3.18) corresponding to Case 3 of Table 1 with $A_w = 0$ has been selected as the basic source model to cover both shallow and deep water, near and far field.

This source model represents an equivalent source displacement $\eta_0(r)$ with zero initial velocity $w_s(r)$ on the surface. Neglecting the initial velocity has little effect on the trailing waves in the case of deep water simulation. In the shallow water case, only the source disturbance $\eta_0(r)$ is primarily responsible for the leading waves. The two governing parameters for this source model are A_η and R_η. For simplicity, they are designated as A and R for the case $w_s(r) = 0$. It is recalled that the transform of the source model gives

$$H_\eta(k) = -\frac{AR}{k} J_3(kR) \,. \tag{3.20}$$

The two correlation parameters, A and R, are a function of yield, water depth, and burst depth. If A and R are specified, the wave elevation at any distance r and time t, $\eta(r, t)$, can be computed utilizing the methods presented in Chapter 2.

4. Calibration

The two parameters, A and R, are a function of yield W, water depth d, and burst depth z. More generally, they are probably a function of ocean bottom or soil characteristics as well, especially for the case of shallow water. The burst depth effect is more evident in deep water than shallow water and will be treated separately. As to the effect of the soil characteristics on the bottom, no sufficient data is available for the present correlation.

The calibration of the mathematical model then involves the correlation between the two mathematical parameters, A and R, and the two explosion parameters, W and d,

$$A = f_1(W, d)$$

$$R = f_2(W, d)\,, \tag{3.21}$$

where A, R, and d each has a linear dimension, and W is normally given in terms of weight.

The correlation of the mathematical equivalents of the disturbance with realistic explosion parameters can be obtained by comparing the predicted wave trains with the measured data of known yield W and water depth d. For any given r and t, the predicted wave train is given in terms of the correlation parameters as follows:

$$\eta(r, t) = -A(W, d)R(W, d) \int_0^\infty J_3[kR(W, d)] J_0(kr) \cos \sigma t \, dk\,. \tag{3.22}$$

This mathematical wave form provides the basis for comparison with experimental wave records of known explosions to calibrate the correlation parameters A and R.

In general, the process of calibration can be divided into two steps; first, the radial parameter R and then the intensity parameter A. The radial parameter R governs the time interval of free-surface oscillations but contributes little to the amplitude of the waves. The general shape, the location of peaks and troughs, and the zero crossings with respect to time, especially that of the leading wave in shallow water or the maximum wave of the wave train in deep water, generally determine the value of R. Once R is determined, the intensity parameter A can be determined by comparing the absolute value of the amplitude obtained from the mathematical model with that obtained from measurements.

The above procedure would be valid given sufficient data. The amount of measured data is nonetheless limited, especially over the range of intermediate water depth. In order to establish a uniformly valid model for either small or large yield and shallow or deep water, incorporation of analytical procedures in the correlation is necessary.

The method is a hybrid of forward and inverse procedures as opposed to strictly using the inverse procedure discussed previously. The hybrid procedure involves a forward approach using a preassumed mathematical model for wave propagations, calibrated by inversely correlated empirical formula of the maximum wave in the wave train.

Recognizing $d/W^{1/3}$ as a valid scaling parameter between water depth and yield, Equation (3.20) can be written in the following dimensionless form

$$\eta^*(r^*, t^*) = -A^*\left(\frac{d}{W^{1/3}}\right)R^*\left(\frac{d}{W^{1/3}}\right)$$
$$\times \int_0^\infty J_3\left[k^* R^*\left(\frac{d}{W^{1/3}}\right)\right]J_0(k^* r^*)\,\cos\,\sigma^* t^* dk^* . \qquad (3.23)$$

The dimensionless variables have been defined by Equation (2.1). In addition:

$$A^* = \frac{A}{d}$$

$$R^* = \frac{R}{d} .$$

The procedure assumes *a priori* that the proposed model (3.23) is valid and calibrates the two unknown parameters, A^* and R^*, by the empirical formula (3.4). One has seen in Chapter 2 that the evaluation of the integral in Equation (3.23) is difficult when r^* is small or $t^* \approx r^*$, and that this integral can be approximately evaluated by the method of stationary phase valid for large r^* and t^* (2.67).

Then alternatively, the integral in Equation (3.23) can be more conveniently evaluated by an inverse Fourier transform as shown in Equations (2.37) to

(2.50). This is purely a mathematical representation subject to the postulation of the source disturbance. Analytically evaluating $\eta(r, t)$ by varying R^* from small to large, the time history of η/A for each R^* at several different r^*s, is computed and shown in Figures 3.10–3.14. From these plots, it is clear that small values of R^* generally yield deep water characteristics of wave propagation while large R^*s give the shallow water wave characteristics. The transformation of wave characteristics from near to far field may be directly observed from the results at different locations of r^* for a given R^*, but the general features of near and far field wave characteristics can also be identified by examining the results for various r^*/R^* values from small to large regardless of the value of R^*. One may also observe from the wave trains that the leading wave generally is the largest wave in shallow water (large R^*) or near field (small r^*/R^*), but the largest wave shifts to the trailing group in the far field (large r^*/R^*) or in deep water (small R^*).

The value of R^* is a measure of the disturbance relative to the water depth. It is anticipated that R^* must be intimately related to wave decay. By plotting the maximum η^* versus r^*, the decay characteristics for different values of R^* are shown in Figure 3.15. The well-organized trends of these results are clearly demonstrated in the figure. The slope of these numerical results on the log-log plot provides the information on the decay of these waves. The data align orderly toward a linear decay or $-45°$ slope when R^* becomes very small. The other extreme is the 2/3 power decay or $-33.69°$ slope when R^* approaches 10 or higher. The solid lines drawn in the plot are to show the variation of decay, corresponding a linear interpolation of slopes between 1 and 2/3 for R^* between 0 and 10, respectively. The slope of the solid lines seems to support the trends of the data points very well.

It must be noted that the points shown in Figure 3.15 are computed results using the mathematical model (3.22) with systematic variations of R^* or R/d. The solid lines are not regression of these points, however. The solid lines are intended to show only the variation of the decay law, should it vary linearly between 1 and 2/3 for R/d varying between 0 and 10, respectively. The computed data points appear not to be smoothly varied in some cases of large R^* as shown in the plot. This is due to a large phase shift of the maximum waves in shallow water explosions (large R^*). The maximum wave is generally the leading wave in the case of large R^*, but it is not necessarily the largest possible wave in the theoretical wave envelope. In deep water explosions or small R^* cases, the maximum waves are likely to coincide or are close to

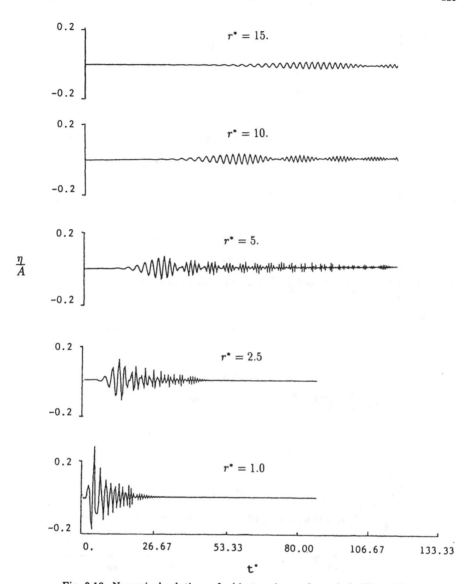

Fig. 3.10. Numerical solutions of η/A at various values of r^*, $R^* = 0.5$.

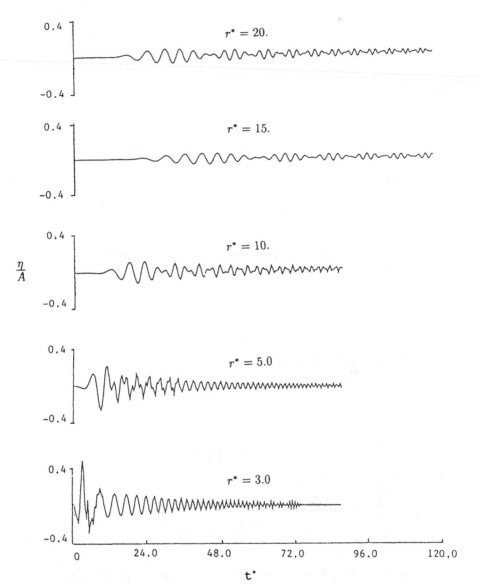

Fig. 3.11. Numerical solutions of η/A at various values of r^*, $R^* = 2.5$.

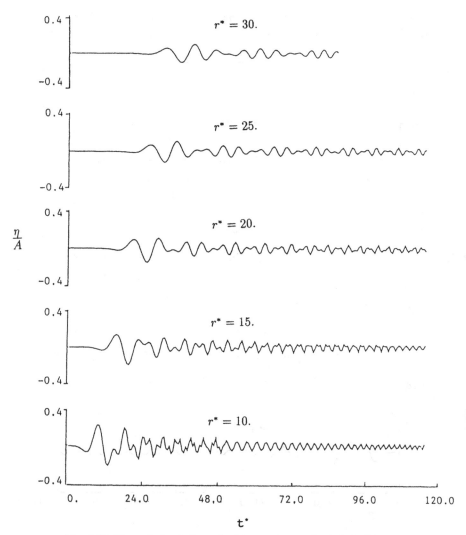

Fig. 3.12. Numerical solutions of η/A at various values of r^*, $R^* = 5$.

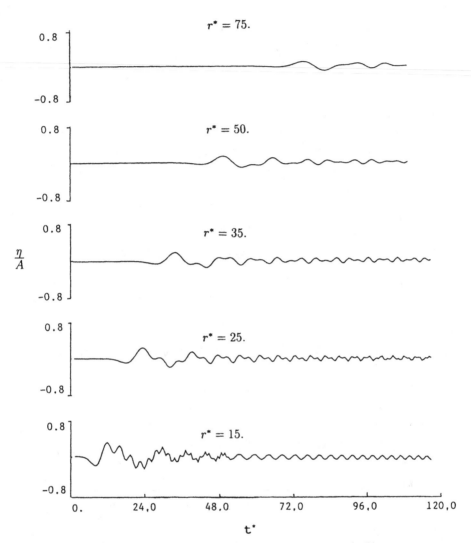

Fig. 3.13. Numerical solutions of η/A at various values of r^*, $R^* = 10$.

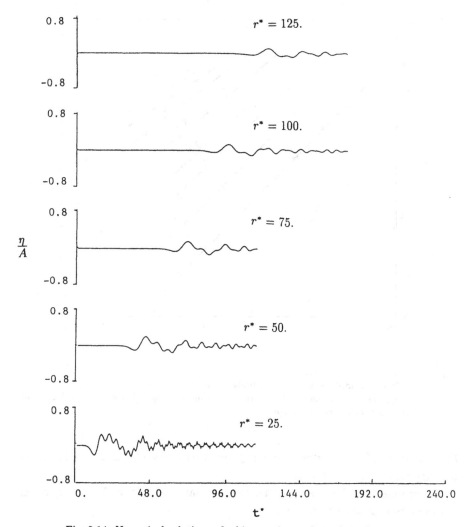

Fig. 3.14. Numerical solutions of η/A at various values of r^*, $R^* = 15$.

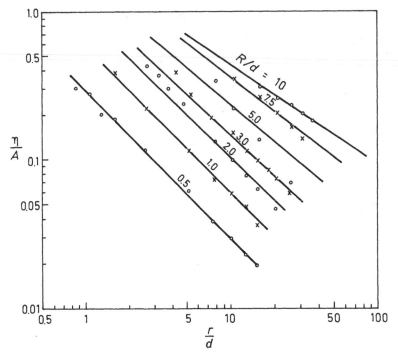

Fig. 3.15. Decay of η as a function of R.

     o & x — computed values from mathematical model 3.22.

     Solid lines — assuming linear interpolation of slopes between 1 and 2/3 for R/d between 0 and 10, respectively.

the maximum of the wave envelope, as there are more waves in a wave envelope in deep-water generated waves.

    Incorporating relationship (3.4), the correlation of R^* with the explosion parameters is derived as follows

$$R^* = 30 - 24.9\left(\frac{d}{W^{1/3}}\right)^{0.07}, \qquad 0.05 < \frac{d}{W^{1/3}} < 14. \qquad (3.24)$$

This relationship suggests that when $d/W^{1/3}$ is small, R approaches $10d$. It also suggests that when $d/W^{1/3}$ is greater than 10, R is less than $0.75d$, sufficient to be considered as a deep water explosion in so far as the definition of R is concerned.

For a given R^*, the value of $d/W^{1/3}$ is determined from Equation (3.24). The empirical formula (3.5) then can be used to estimate, for the particular explosion $(d/W^{1/3})$, the maximum wave or the leading wave, which compares with that of the computed wave train of the corresponding R^* to determine the parameter A^*. A calibration of the parameter A^* with the explosion parameters is given by

$$A^* = 2.7(R^*)^{-0.73}\frac{W^{1/4}}{d}, \qquad 0.05 < \frac{d}{W^{1/3}} < 14. \qquad (3.25)$$

In the 1986 WES tests, wave records for five different yields (10, 20, 30, 40 and 50 pounds) in shallow water, 2.7 feet deep, were collected at several locations from the origin. The natural mud bottom was approximately flat. The charge position was at half-depth. Using Equations (3.24) and (3.25), the calibrated A^* and R^* for the five shots are listed in the following:

W (lbs)	A^*	R^*
10	0.572	4.71
20	0.639	5.12
30	0.690	5.34
40	0.720	5.51
50	0.750	5.64

Computed time histories at selected locations using mathematical model (3.22) with the above calibrated parameters are compared with the actual recordings (Figures 3.16–3.20). The similarities in both magnitude and the major phase shifts are remarkable, despite the crudeness and simplicity of the model.

A comparison with the 1988 WES test of a 10-pound yield in 4 feet of water, on the other hand, is not as good (see Figure 3.21). The calibrated A^* and R^* from Equations (3.24) and (3.25) are 0.435 and 4.0 respectively. The difference between the 1986 and 1988 tests is the bottom of the water body. The 1988 test was conducted in water over a concrete slab, instead of the mud bottom used in the 1986 tests. The bottom soil structure therefore seems to have a significant influence in shallow water explosions. Unfortunately, there is insufficient data to support bottom soil calibrations at the present time.

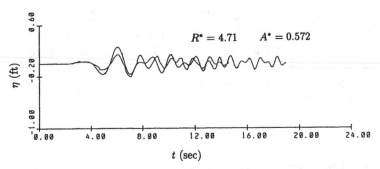

Fig. 3.16. Comparison of the computed result with measurement at $r = 42.5$ feet, for a 10-pound shot in 2.7 feet of water.

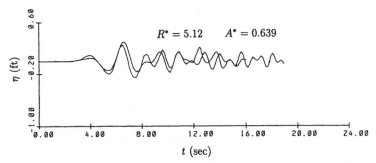

Fig. 3.17. Comparison of the computed result with measurement at $r = 47.5$ feet, for a 20-pound shot in 2.7 feet of water.

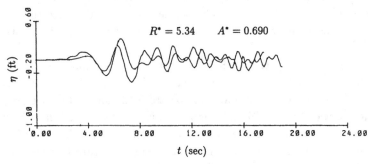

Fig. 3.18. Comparison of the computed result with measurement at $r = 47.5$ feet, for a 30-pound shot in 2.7 feet of water.

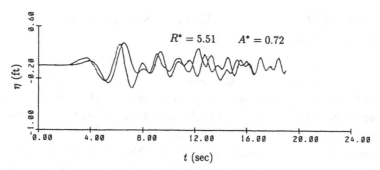

Fig. 3.19. Comparison of the computed result with measurement at $r = 47.5$ feet, for a 40-pound shot in 2.7 feet of water.

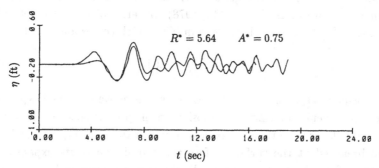

Fig. 3.20. Comparison of the computed result with measurement at $r = 52.5$ feet, for a 50-pound shot in 2.7 feet of water.

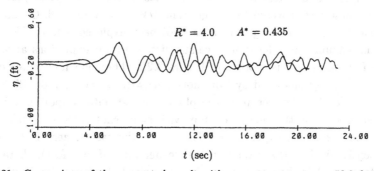

Fig. 3.21. Comparison of the computed result with measurement at $r = 50.2$ feet, for a 10-pound shot in 4 feet of water.

5. Depth of Burst (DOB) Effect, Postulation for Subsurface Explosions

The effect due to the depth of burst has been observed in the results of deep water explosion experiments (Pace *et al.*, 1970). Surface waves were observed to be particularly amplified at two different DOBs, which are known as the upper and lower critical depths. Deep water correlations are customarily prescribed by two different sets of relationships.

In the shallow water experiments (WES, 1955), the DOB effect is not evident, except in a thin layer very near the surface. The observed waves are substantially lower for explosions detonated in this thin layer beneath the surface. Allen (1979) analyzed a series of shallow water tests conducted by URS (Kaplan and Cramer, 1963) together with the Mono Lake tests (Whalin, 1966) and some nuclear events (DNA EM1, 1978). In terms of W, Allen defined this layer as $0.5 \, W^{1/3}$ meters, where W is in KT, which corresponds to the range

$$0 < \frac{z}{W^{1/3}} < 0.013 \tag{3.26}$$

in the foot-pound system. It is a very thin layer; even well-planned experiments would have difficulty placing the explosives in the precise position. In addition, because this layer is so thin, an explosion within this thin layer must have its bubble breached at the surface before the completion of its expansion. In order to analyze the problem more reliably, any explosion within this layer is excluded from being considered as a fully contained subsurface event.

The 1988 WES tests were 10-pound explosions at several DOBs in water of 2.7 feet and 4 feet in depth. In terms of the depth-yield parameter $d/W^{1/3}$, these tests covered the range from $d/W^{1/3} = 1.25$–1.86, the intermediate water depth, the range which had not been explored before. The decay of the maximum wave elevation η versus distance r for explosions at various levels of submergence is plotted in Figure 3.22. The solid line indicates the anticipated decay predicted by Equation (3.4) with $\alpha = 0.84$ and 0.87 for the two water depths. As seen from the plots, the test data support the general trend suggested by the prediction but with significant scatter, which is not unusual for explosion experiments. Except for the case of zero DOB, which consistently yields a lower wave in the deeper case ($d = 4$ feet), there is no consistent trend leading to any conclusion toward the burst depth effect. If such data are plotted as a function of the DOB ratio (z/d), the same degree of

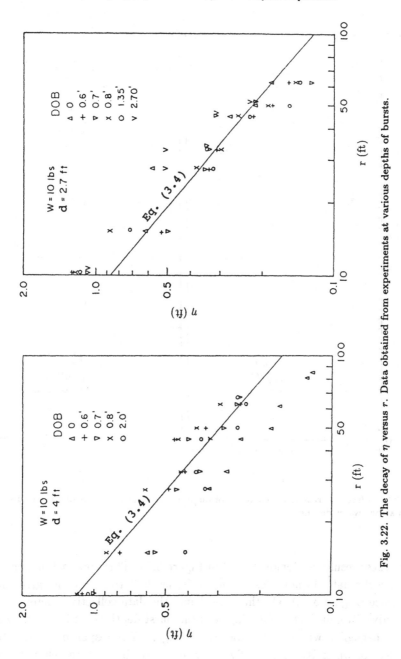

Fig. 3.22. The decay of η versus r. Data obtained from experiments at various depths of bursts.

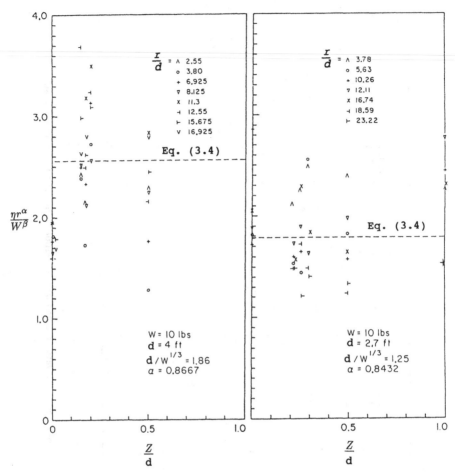

Fig. 3.23. Wave generation versus depth of burst *z*. Experimental data of 10-pound explosions at two water depths.

scatterness results as demonstrated in Figure 3.23. This approach is unsound because the data do not present any self-consistent trend for any given submergence (Figure 3.22). On the other hand, the data collectively support the predictive formula (3.4) indicating that the burst depth effect is probably minor or negligible within a certain error margin. The response of zero DOB in 4 feet of water is substantially lower, evidently because the charge is too close to the surface and should not be categorized as subsurface explosions as inferred above.

It is clear that the decreasing burst depth effect in intermediate and shallow water cannot be generalized to deep water, as there is strong evidence that waves are particularly pronounced due to bursts at the upper critical depth as explained in Chapter 1. Besides the pronounced amplitude peak at the upper critical depth, data also show a mild hump at the so-called lower critical depth (Figure 1.8). As Whalin *et al.* (1970) indicated however, this lower critical depth cannot be defined, if the data of very small charges are discarded (Figures 3.24 and 3.25). On the other hand, the peak at the upper critical depth seems to definitely exist for either large or small charges. This indicates that the burst depth effect is governed not simply by the reduced submergence (submergence depth of burst/linear scale reduced from yield). It seems evident that the optimal submergence should be related to the size of the bubble formed by the explosion. The size of the bubble is a function of the yield but the optimum position of the bubble for wave generation depends on the water depth, if the water is not sufficiently deep. This leads to a parameter which should be a function of both $z/W^{1/3}$ and z/d, or alternatively written as

$$\zeta = f\left(\frac{z}{W^{1/3}}, \frac{d}{W^{1/3}}\right). \tag{3.27}$$

The charge submergence depth governs the relative importance of the pressure and gravity effects during the process of bubble expansion. In an analysis for the optimal burst depth, Schmidt and Housen (1987) selected a parameter of the following form

$$p_o = 1 + \frac{KP}{\rho g z} \tag{3.28}$$

to account for the relative measure between the atmospheric and the static overburden pressure as seen in Figure 1.8. In Equation (3.28), K is a constant determining the transition between the pressure-predominated and gravity-dominated regimes. Schmidt and Housen adopted a value, $K = 10$, based on the scaling of soil cratering but additionally stated that the value of K is not critical. To account for the submergence depth effect, a form similar to Equation (3.28) is considered here, but the product KP is taken instead as a measure of the bubble pressure. Since this pressure should be a direct measure

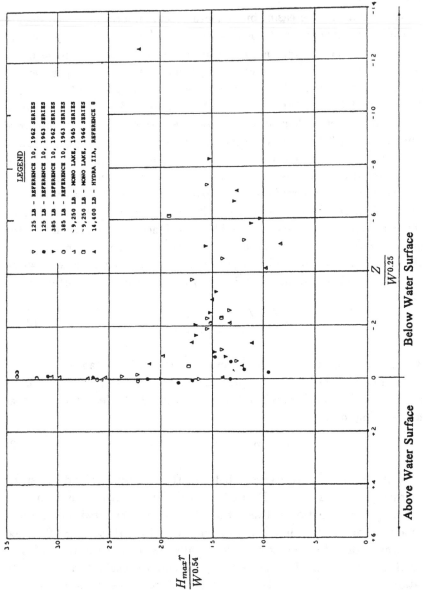

Fig. 3.24. Deep water wave data for charge weights greater than 100 pounds (from Whalin *et al.*, 1970).

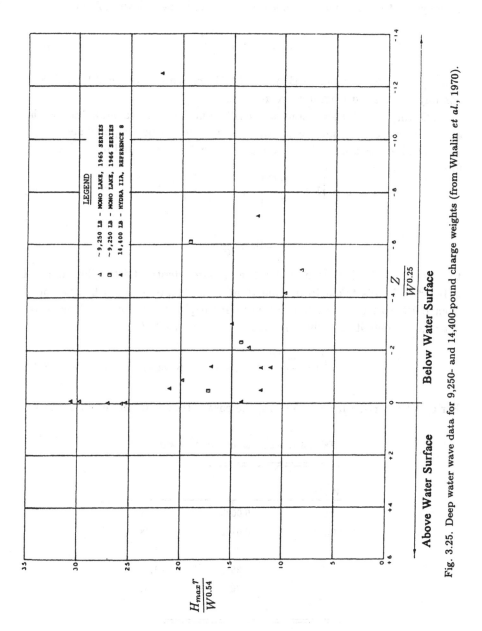

Fig. 3.25. Deep water wave data for 9,250- and 14,400-pound charge weights (from Whalin et al., 1970).

of the yield, the parameter p_0 is assumed to be in the following form

$$p_o = 1 + c_1 \frac{W^{1/3}}{z} \qquad (3.29)$$

in the present correlation. In Equation (3.29), c_1 is a proportionality constant, possibly determined from data correlation.

Following Schmidt and Housen's result from data regression, assuming the generation process of surface waves in deep water to be proportional to $p_o^{0.2}$ and tentatively assigning $c_1 = 1$, the parameter ζ is postulated in the following form

$$\zeta = \left[1 + \frac{W^{1/3}}{z}\right]^{c_2}, \qquad \frac{z}{W^{1/3}} > 0.05, \qquad (3.30)$$

with

$$c_2 = 0.014 \frac{d}{W^{1/3}}, \qquad (3.31)$$

to account for both the burst depth and water depth effects. The limit of $z/W^{1/3} > 0.05$ is taken to be consistent with Equation (3.26) but extended from 0.013 to 0.05 to exclude near-surface explosions. For the deep water extreme, by definition defined in Equation (3.15), $d/W^{1/3} = 14.32$,

$$\zeta = \left(1 + \frac{W^{1/3}}{z}\right)^{0.2}, \qquad \text{(deep water)}. \qquad (3.32)$$

For $z/W^{1/3}$ varying from 0.05 to 14, the parameter ζ is tabulated in Table 3.1.

Table 3.1. Variation of ζ in deep water.

$\dfrac{z}{W^{1/3}}$	$\dfrac{z}{d}$	ζ
0.05	0.0035	1.84
0.25	0.0175	1.38
1	0.07	1.15
5	0.35	1.04
10	0.70	1.02
14	~ 1	~ 1

The above postulation seems to support the existence of an upper critical depth. The pronounced amplification also seems to be limited to the layer of $0.05 < z/W^{1/3} < 0.25$ and gives a factor of 30–80% higher than that for deeply submerged cases. It should be noted, however, the postulation of the factor ζ (3.30) at the present stage is purely a conjecture which needs more credible evidence to prove or calibrate. As a minimum, the proportionality constant c_1 in Equation (3.29) and the exponent c_2 in Equation (3.31) should be carefully correlated using systematically established data, which are not available at the present time.

Also note that the present analysis is confined to subsurface explosions, which are tentatively defined by those depths of the burst z greater than a critical depth z_c $(= 0.05W^{1/3})$ as indicated in Equation (3.30). It is believed that this limit is related to the depth at which the bubble may reach its maximum expansion for that submergence before breaching the surface, and consequently this depth may be intimately related to the so-called upper critical depth. Undoubtedly, this critical depth is a function of the yield.

In shallow water where $d/W^{1/3} < 1$, the parameter ζ approaches unity with little influence from charge location z, which agrees with the evidence shown in the shallow water test data. In terms of z/d and $d/W^{1/3}$, the parameter ζ is tabulated in Table 3.2.

Table 3.2. Variation of ζ as a function of z/d and $d/W^{1/3}$.

$\dfrac{z}{d}$	$\dfrac{d}{W^{1/3}}$					
	0.1	0.5	1.0	2.0	6.0	10.0
0.005						1.53
0.01					1.27	1.40
0.05			1.04	1.07	1.13	1.17
0.1		1.02	1.03	1.05	1.09	1.10
0.25		1.02	1.02	1.03	1.04	1.05
0.5	1	1.01	1.02	1.02	1.02	1.03
0.75	1	1.01	1.02	1.02	1.02	1.02
1	1	1	1	1	1	1

6. Correlation of Burst Depth Effect in the Mathematical Model

To account for the DOB effect, the following parameter is postulated for empirical correlations:

$$\Lambda = \frac{\zeta}{\zeta_c} = \left[\frac{1 + \dfrac{W^{1/3}}{z}}{1 + \dfrac{W^{1/3}}{z_c}}\right]^{c_2} , \qquad z \geq z_c . \tag{3.33}$$

When z_c has been assigned a value equal to $0.05\, W^{1/3}$,

$$\Lambda = \left[0.0476\left(1 + \frac{W^{1/3}}{z}\right)\right]^{c_2} . \tag{3.34}$$

The parameter Λ may be regarded as a factor to incorporate various burst levels to the response.

Further note that the propagation of the surface waves and the energy coupling depend on the charge submergence depth as shown from the test data of Pace *et al.* (1970). This indicates that wave propagation itself is a function of $z/W^{1/3}$ in addition to $d/W^{1/3}$. To include the DOB effect, therefore, a new parameter $\mathcal{D}$ defined by

$$\mathcal{D} = \Lambda \frac{d}{W^{1/3}} \tag{3.35}$$

is used to replace $d/W^{1/3}$ in the correlation relations. Again, when $d/W^{1/3} > 14$, it is taken as equal to 14 so that $\mathcal{D} \leq 14\Lambda$.

Consequently, the empirical formula (3.4) can be written as follows:

$$\frac{\eta_{\max} r^\alpha}{W^\beta} = \Lambda C_W , \tag{3.36}$$

or

$$\frac{\eta_{\max} r^\alpha}{W^\beta \Lambda} = C_W , \tag{3.37}$$

where

$$\alpha = 0.83 \mathcal{D}^{0.07}$$

$$\beta = \frac{\alpha}{3} + 0.25 \tag{3.38}$$

$$C_W = 1.44 \mathcal{D}^{0.93} .$$

In shallow water when $d/W^{1/3} \leq 1$, $\Lambda \to 1$, and $\mathcal{D} \to d/W^{1/3}$, Equation (3.37) reduces to Equation (3.4) identically.

To be consistent with Equation (3.37), the correlation parameters R (3.24) and A (3.25) are modified and given by

$$R = (30 - 24.9\mathcal{D}^{0.07})d, \qquad\qquad (3.39)$$

$$A = 2.7\Lambda^{1+\alpha}\left(\frac{R}{d}\right)^{-0.73} W^{1/4}. \qquad\qquad (3.40)$$

Again, in Equations (3.37), (3.39), and (3.40), all the linear dimensions are in feet and yield W in pounds. This correlation indicates that the depth of submergence is directly related to the behavior of wave propagation. The crater radius becomes larger, the waves become longer but smaller, and waves decay more slowly when the charge submergence moves toward the lower level. The correlation of R (3.39) is not anticipated to be valid for deep submergence beyond a certain critical level. However, the effect of charge submergence becomes negligible in shallower water when $d/W^{1/3} \leq 1$.

Following the empirical relationship of Equation (3.37), all experimental (nonnuclear) data are plotted and shown in Figure 3.26. The data cover test points for $0.05 < z/W^{1/3} < 4$. By plotting the experimental values of $\eta_{\max}r^\alpha/\Lambda W^\beta$ versus $\mathcal{D}$, the data are shown to be closely aligned with the empirical line of $C = 1.44\mathcal{D}^{0.93}$. The scatter of the data exhibited in Figure 3.23 essentially reflects the scatter of the actual event, which is especially evident for the upper critical depth cases as they are intrinsically unstable. Consequently, regardless of being an upper or lower critical event or at other submergence levels, the empirical prediction from Equations (3.37) and (3.38) should provide a guide for an event of a given water depth ($d/W^{1/3}$) and submergence ($z/W^{1/3}$).

7. Wave Making Efficiency

Theoretically, the energy in wave formation from an underwater explosion can be estimated from the potential and kinetic energies of the initial disturbance postulated in the mathematical model. The present model takes a static crater $\eta_0(r)$ described by Equation (3.18). Because no initial velocity is assumed in the model ($w_s(r) = 0$), only potential energy contributes to

WES (1955)   × W = 0.5-2048 lbs
WES (1961)   □ W = 0.5 lbs
             ⋇ W = 2 lbs
             ▲ W = 10 lbs
             + W = 125 lbs
             ∗ W = 385 lbs
WES (1986)   ◊ W = 10-50 lbs
WES (1988)   ○ W = 10 lbs
WES (1989)   ∗ W = 0.5-10 lbs
MONO LAKE (1966) ◦ W = 9250 lbs

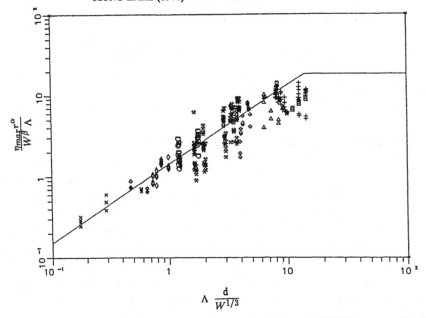

Fig. 3.26. Empirical fit of maximum waves by generalized formula (3.37) in terms of yield, water depth, submergence depth, and distance.

wave formation. The potential energy for a static crater (1.10) is given by

$$E_w = \frac{\pi}{6}\rho g A^2 R^2 \tag{3.41}$$

with A and R defined in Equation (3.18) as A_η and R_η. Assuming the foregoing correlations are valid and substituting in Equation (3.41) the relationships of A and R from Equations (3.39) and (3.40), the energy of wave formation can be expressed as a function of the parameter $\mathcal{D}$, which is in turn a function of yield W, water depth d, and charge submergence depth z. On the basis of one pound of TNT releasing 5×10^5 calories, the efficiency of wave generation from an underwater TNT explosion is given by

$$\varepsilon = \frac{E_w}{E} = 158.35 \times 10^{-6} \Lambda^{2\alpha} \mathcal{D}^2 (30 - 24.9 \mathcal{D}^{0.07})^{0.54} W^{1/6} . \qquad (3.42)$$

Equation (3.42) indicates that if $\mathcal{D}$ is kept constant, wave generation efficiency increases as $W^{1/6}$, where W is the weight of TNT in pounds. Efficiency increases also with water depth. In water of identical depth, the lower yield explosion is more efficient than the higher yield. The generation efficiency for a mid-depth burst is given in Table 3.3. In the case where these figures can be compared with the wave making efficiency obtained by using the nonlinear wave theory as given in Chapter 5, the discrepancies should lead to the conclusion that the linear wave theory is inadequate in shallow water. Also,

Table 3.3. Wave generation efficiency for burst at mid-depth submergence (d in feet, W in KT).

$\dfrac{d}{W^{1/3}}$	W (KT)	d (ft)	ε(%)
	1	126	.37
1.0	10	271	.54
	100	584	.80
	1000	1259	1.17
	1	252	1.10
2.0	10	542	1.61
	100	1169	2.36
	1000	2519	3.47
	1	630	3.25
5.0	10	1357	4.77
	100	2924	7.01
	1000	6299	10.3
	1	1259	4.37
10.0	10	2714	6.41
	100	5848	9.41
	1000	12599	13.80

in deep water, the large increase of efficiency exhibited in Table 3.3 indicates that there are limits of validity even though ill-defined, as much less energy is being found in water waves than is given by Equation (3.42). The experimental evidence on which the empirical formulae are based simply does not exist, and one should be careful in extrapolating the range of application of the formula presented in this chapter beyond the range of experiments on which they are based.

8. Summary

This chapter presented a single mathematical model to describe the wave history due to an underwater explosion at any level of submergence and a single empirical formula to predict the maximum wave amplitude in water of any depth. The nearest distance applies to the smallest circle where the leading wave feature is distinctly important but the nonlinear phenomenon can be ignored; it is generally three to four plume radii outside the source center. With this understanding, the model and predictive formula are uniquely valid for near and far field, shallow and deep water, shallow and deep burst, underwater explosions over a uniform depth bottom.

The subsurface event is defined by limiting the burst depth to be under a certain critical layer, which is contemplated as a depth at which the bubble of the expansion may reach its maximum radius for that submergence before breaching the surface. This depth is believed to be intimately related to the so-called critical depths, although no exact relationship has been established.

The burst depth, as well as the water depth and yield, affects the prediction through a definitve relationship. Since 30–40% or even 100% discrepancy in observations from supposedly identical explosions is common, the uncertainty exhibited in the prediction model can be explained by the actual uncertainty of the physical process.

The model provides an additional dimension to examine various effects due to the burst depth; it directly affects the crater (or plume) radius, the wave period, as well as the wave amplitude. In some sense, an event of a deeper burst depth seems to be equivalent to a relatively shallower water event. The effect of burst depth on very shallow water events, however, is negligible.

The water depth versus the charge yield ratio is an important parameter even in "deep water" explosions, because no ocean is really deep for yields of more than one megaton. Consequently, the efficiency in wave formation of an underwater explosion does not increase with yield as anticipated in the past, instead it decreases for larger yield in water of equal depth. For a given yield, the efficiency increases rapidly in deeper water. However, one must be careful when applying the empirical relationship presented in this chapter beyond the range of experiments on which they are based.

Chapter 4

NONLINEAR WAVE GENERATION MECHANISM AND WAVE PROPAGATION IN SHALLOW WATER

1. Physical Overview and Theoretical Approach

The phenomena described in the following two chapters apply when the relative depth $(d/W^{1/3}) < 1$. Then the water crater is considered to reach the seafloor and the nonlinear effects are of such importance that they can no longer be neglected as in the two previous chapters. The development of the corresponding theories has been done recently. The contribution of Khangaonkar (1990) to the solution of these problems has been tremendous, allowing a quantum step in the understanding of shallow water explosion. Most of the material presented in this chapter is from his work in a series of publications which includes Khangaonkar (1990), Khangaonkar *et al.* (1991), and Le Méhauté *et al.* (1992).

The distinguishing characteristics of EGWWs in shallow water are (see Figure 1.3):

(1) A large leading wave which behaves almost like a solitary wave, followed by an extended long trough.

(2) The long trough is followed by a succession of conoidal-type waves of increasing frequency akin to an undular, nondissipative bore.
(3) The wave-trough conoidal-type system is followed by a group of higher frequency waves which are of Stokesian type.
(4) At large "t" (period of propagation) and large distance r from GZ, as the wave amplitude decreases, the waves are of the linear Airy type.

The linear model is not capable of generating a wave record bringing out these features. A number of small scale tests (Waterways Experiment Station, 1988 and 1989) have been done to simulate shallow water EGWWs. After studying the video tapes of the tests it became apparent that the shallow water phenomenon could be split into a number of small phenomena which could be analyzed in sequence.

A first look at the videos shows a seeming chaos following the explosive burst and the splash of water resulting in a lot of turbulent agitation. A careful observer can however discern the pattern contained in every test case that is pertinent and responsible for water wave generation. Most of the explosions (1 pound to 50 pounds of TNT) in 2.7 feet of water created a crater that reached the bottom and exposed it. The body of water beyond the crater seemed to be undisturbed, while the charge released pressure to the atmosphere during its first expansion phase, creating a large vertical plume and a hollow crater. The expansion of the crater created a lip which evolved into a cylindrical breaker, which is damped rapidly to form the leading wave. Meanwhile, the cylindrical crater collapsed inward on the dry bed and water converged at the center resulting in a reflection on itself. The water accumulated in the center, the resulting column rose and began to fall while dissipating energy very similar to that in a moving hydraulic jump. This propagated a short distance like a second bore. As this bore height diminished and the flow became tranquil, the dissipative phenomenon stopped and a high frequency trailing wave group was observed well behind the leading wave, which had a considerable head start over the rest of the waves behind it. The waves propagated in the form of an undulated bore in the region where the amplitudes were large in comparison to the water depth. Thus, the waves retained their nonlinear structure of steep crests and shallow troughs as they propagated outwards. Meanwhile, the collapse of the plume on the forming waves, adding considerable water to the moving wave, caused some damping by momentum transfer and turbulence. After traveling a certain distance due to cylindrical spreading and dispersion, the wave amplitudes became small, the waves became linear and of the Airy type.

Study of several tests in slow motion enabled us to decipher the complex phenomenon of shallow water wave generation and propagation described above (see Figure 1.4). Based on our understanding and evaluation of the physics of the phenomenon, the following specific objectives are laid out.

(1) Using a theoretical crater with a lip as the initial free-surface disturbance with the bed exposed to obtain a solution of the hydrodynamics of the collapse of a cylindrical cavity on a dry bed.

(2) A solution for the hydrodynamic dissipative losses in the bore formation from the self convergent crater collapse and its propagation behind the leading wave.

(3) An evolution of the dissipative bore and the leading wave into nonlinear nondissipative Boussinesq or Conoidal waves.

(4) Development of a technique to propagate waves from weakly nonlinear dispersive region into fully dispersive linear region.

(5) Preliminary calibration of the nonlinear shallow water model using the near field test data. Estimation of efficiency based on the preliminary calibration.

2. Wave Generation Mechanism

We begin with the assumption that in shallow water $(d/W^{1/3} \leq 1)$, the underwater explosion crater radius is much greater than the water depth. The force of the explosion pushes the water outwards creating an axisymmetric cavity extending down to the bottom and exposing it almost entirely. The pressure meanwhile is released to the atmosphere from the free surface. The displaced water is prevented from moving outwards by the presence of a large body of stationary water whose inertia cannot be overcome by the radial momentum of a relatively small volume of water from the cavity. Thus, most of the water is thrown upwards creating a large plume visible on the surface. Figure 4.1 shows the crater and the plume formation schematically due to a shallow water explosion. Water particles beyond the crater and the plume are observed to be calm and relatively unaffected by the burst. The shock wave effects do not interfere with the process of water wave formation and may be neglected.

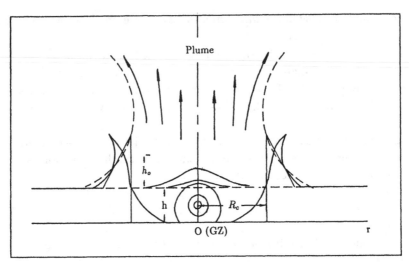

Fig. 4.1. Expansion of the explosion bubble and its subsequent burst through the free surface and theoretical representation of the physical crater with a cylindrical cavity surrounded by a lip. h is the water depth, h_0 is the height of the lip and R is the crater radius.

Based on the scenario described above and taking into account the mathematical convenience, the crater and the plume may be replaced with a cylindrical static cavity and a lip as a theoretical model for the initial disturbance to the body of water (Figure 4.1). The size of the static theoretical cavity is very close to the maximum size of the physical cavity due to the explosion, when most of the kinetic energy is nil.

The lower part of the plume without the splash of water, is equivalently represented by the release of a volume of water above the free-surface level on the perimeter of the cavity. This results in the leading wave just as in a real explosion. The plume falling under gravity may also contribute to the formation of the leading wave (based on observation). The dimensions of the lip are governed by the amplitude of the leading wave and conservation of mass (mass lost in the crater = mass of water in the lip).

The exact shape of the crater and the lip can be obtained using Hydrocodes. But the state of the art for application to explosion-generated craters is still inadequate for generalized usage. Thus, for the purpose of mathematical modeling of the wave generation phenomena, a cylindrical cavity with a lip has been chosen to represent the initial conditions. While it is obvious that the real "crater-plume" system produced by an explosion would not be as well defined as its mathematical equivalent, it is our contention that the acute geometrical features like the nonuniform crater surface, exact shape of the lip, etc. will not affect the large scale wave effects. Dispersion coupled with cylindrical spreading will quickly smooth out the effects of small scale sharp discontinuities, and generate water waves such that important wave parameters, such as the wave height, wave period, and the group formations, will be well reproduced. In other words, waves resulting from the exact explosion-generated crater and that resulting from its mathematical fit would be equivalent for all practical purposes.

Thus the problem of wave generation by underwater explosions in shallow water is formulated as, "The Waves Caused by the Collapse of a Cylindrical Cavity with a Lip onto a Dry Bed". The parameters which govern the wave generation, namely, crater radius and the lip size and shape are the unknowns which will be determined by calibration with the experimental data (Chapter 5). The collapse of the cavity is followed by the formation of a dissipative bore resulting from the reflection on itself at the center of the cavity or at GZ. This bore follows the leading wave up to the point where the flow becomes tranquil and the bore is converted to a group of trailing waves.

3. Collapse of a Cylindrical Cavity onto a Dry Bed — Governing Equations

The collapse of a cylindrical cavity onto a dry bed can be analyzed analogously to a "Dam-Break" problem in cylindrical coordinates. On releasing the initial cavity, the leading edge on the lower side of the "Dam" may be expected to converge to the center with a splash, while the rarefaction wave travels away causing a drop in the water level as it moves. The release of an initial wall of water has been investigated using the nonlinear long-wave equations in cartesian coordinates.

The nonlinear long-wave equations in cylindrical coordinates in the case of an axisymmetric motion are given by the momentum balance (all notations are nondimensionalized with respect to the still-water depth d, which is labeled h^* in Chapters 4 and 5):

$$\frac{\partial \eta}{\partial t} + \frac{1}{r}[(\eta + h)U] + \frac{\partial}{\partial r}[(\eta + h)U] = 0, \qquad (4.1)$$

and continuity

$$\frac{\partial U}{\partial t} + U\frac{\partial U}{\partial r} + g\frac{\partial \eta}{\partial r} = 0. \tag{4.2}$$

The pressure is hydrostatic. U represents the nondimensionalized, vertically averaged, horizontal velocity in the radial direction. (Note that $h = h^*/d = 1$.)

A classical solution to the dam-break problem in 1D (one-dimensional (x, z, t) coordinate system) has been presented in an analytical form by Stoker (1957). Numerical solutions of the long-wave equations for the dam-break problem have been found using the method of characteristics and the finite difference method (e.g., Katopodes and Strelkoff, 1978; Sakkas and Strelkoff, 1973; Fennema and Chaudhury, 1987).

The method of characteristics is adopted to solve the problem of cavity collapse because of its flexibility in handling shocks and discontinuities and its ability to provide stable solutions.

Application of the method of characteristics involves conversion of Equations (4.1) and (4.2) into a suitable form by performing a change of variables. Wave elevation η is replaced by introducing the quantity C defined by,

$$C^2 = g(h + \eta), \tag{4.3}$$

such that Equations (4.1) and (4.2) can be rewritten as

$$2C_t + 2UC_r + CU_r = -\frac{UC}{r}, \tag{4.4}$$

$$U_t + UU_r + 2CC_r = 0. \tag{4.5}$$

Adding and subtracting Equations (4.4) and (4.5) the following pair of characteristic equations are obtained:

$$\left[\frac{\partial}{\partial t} + (U \mp C)\frac{\partial}{\partial r}\right](U \mp 2C) = \pm\frac{UC}{r}. \tag{4.6}$$

This implies that along the characteristic lines given by

$$\frac{\partial r}{\partial t} = U \mp C, \tag{4.7}$$

the equation

$$\frac{D}{Dr}(U \mp 2C) = \pm\frac{UC}{r} \tag{4.8}$$

holds good.

This is an initial value problem where the free-surface elevation at time $t = 0$, is given [see Figure 4.2(a)]. The characteristic grid is constructed on an $(r - t)$ plane, which means that the r axis contains points from the initial condition. Each point carries with it information on quantities U and C. In the 1D problem, the quantities $U + 2C$ and $U - 2C$ are constant along each characteristic and are referred to as the "Reiman Invariants". However, in the context of the 2D problem in the cylindrical coordinate system, the quantities are no longer invariant. So we will refer to them as "Pseudo Reiman Invariants".

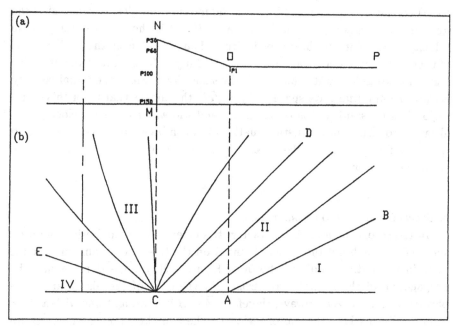

Fig. 4.2.   (a) PONM is the initial free surface.   (b) Schematic representation of the +ve characteristics expected from the collapse of the crater PONM.

Two characteristic lines may be drawn through each point on the r axis. The curves satisfying Equation (4.7) are called C+ characteristics or C− characteristics depending on the sign. Equation (4.8) is integrated along C+ and C− such that at the point of intersection of C+ and C−, the Pseudo Reiman Invariants are fully determined. Thus, four unknowns, U, C, r and t, are solved using the four Equations (4.7) and (4.8) progressively along the characteristic mesh.

Before the application of the method, it is necessary to have a good idea about the expected behavior of the solution and the characteristic mesh. Figure 4.2(b) shows the type of characteristic behavior that might be expected from the initial condition chosen for the solution of the cavity collapse problem. The characteristic mesh is divided into four regions and the solution in each region is discussed below.

The first region is the calm region. A point $p(i, j)$ represents the point p on ith characteristic at jth time step. The location of the point is $r(i, j)$ and $t(i, j)$ on the $(r - t)$ plane and carries the information $C(i, j)$ concerning the free-surface elevation [Equation (4.3)] and $U(i, j)$ is the radial water particle velocity. Any point p in this region is situated far enough from the origin so that at time t, the rarefaction wave has not reached it yet. The characteristics in this region are straight parallel lines with the same slope. Since the initial velocity is $U_0 = 0$ and initial wave speed is $C_0 = \sqrt{gh}$, this means that C for this entire region is a constant and is equal to C_0, and the water particle velocity U is always zero. The characteristic equations in cylindrical coordinates reduce to the same form as in the plane case. This also implies that the free-surface elevation given by

$$\eta = \frac{C^2}{g} - h \,, \tag{4.9}$$

is always equal to zero in this region.

In the second region, the characteristics emerge from the initial condition specified by the lip of the crater. Point A on the $(r - t)$ diagram corresponds to O in the initial condition shown in Figure 4.2(a). $\overline{A - B}$ represents the propagation of the leading wave front. At the edge of the disturbance, the particle velocity is zero always, therefore, $\overline{A - B}$ is a straight line with a slope equal to C_0, marking the end of region I. The boundary points on $\overline{A - B}$ are given by

$$U(1, j + 1) = 0$$

$$C(1, j + 1) = C_0$$

$$r(1, j + 1) = r(1, j) + \Delta r$$

$$t(1, j + 1) = t(i, j) + \frac{\Delta r}{C_0} \,,$$

where Δr is the specified increment along the radial direction. To the left of $\overline{A - B}$, C+ characteristics start off, each with an increasing slope given by

$$\frac{\partial r}{\partial t} = U(i, 1) + C(i, 1). \tag{4.10}$$

The initial conditions here are

$$U(i, 1) = 0$$

$$C(i, 1) = \sqrt{g(h + \eta(i, 1))}$$

$$r(i, 1) = R_{\max} - \nabla r \cdot (i - 1)$$

$$t(i, 1) = 0.$$

$\eta(i, 1)$ is the initial surface elevation of the lip at $r(i, 1)$ and $R_{\max}$ is the distance of the tip of the lip from the origin GZ. Thus, known values of U and C on the line $\overline{A - B}$ are made use of to solve Equation (4.8) along each characteristic to generate the solution and the next characteristic $\overline{A^1 - B^1}$.

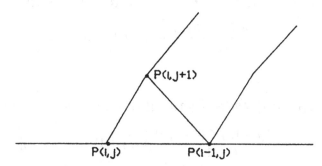

Fig. 4.3. Grid point computation (region II).

As shown in Figure 4.3, the known boundary values of the variables C, U, r and t at $P(i - 1, j)$ along $\overline{A - B}$ and the initial value $P(i, j)$ are used to compute $P(i, j + 1)$. First, the location of the new point $P(i, j + 1)$ is determined. Equation (4.7) is discretized as

$$r(i, j + 1) - r(i, j) = [U(i, j) + C(i, j)][t(i, j + 1) - t(i, j)], \tag{4.11}$$

and

$$r(i,\,j+1)-r(i-1,\,j) = [U(i-1,\,j)-C(i-1,\,j)][t(i,\,j+1)-t(i-1,\,j)]. \quad (4.12)$$

Solving the above two equations simultaneously, we obtain

$$
\begin{aligned}
t(i,\,j+1) = \{&r(i-1,\,j) - r(i,\,j) + t(i,\,j)[U(i,\,j) + C(i,\,j)] \\
&- t(i-1,\,j)[U(i-1,\,j) - C(i-1,\,j)]\}/ \\
&[C(i,\,j) + U(i,\,j) - U(i-1,\,j) + C(i-1,\,j)],
\end{aligned}
\quad (4.13)
$$

and

$$r(i,\,j+1) = r(i-1,\,j) + [U(i-1,\,j) - C(i-1,\,j)] \cdot [t(i,\,j+1) - t(i-1,\,j)]. \quad (4.14)$$

Thus, the new grid point $P(i,\,j+1)$ is generated.

Similarly, to determine the velocity and elevation at the new grid point, Equation (4.8) is discretized as follows:

$$
\begin{aligned}
&[U(i,\,j+1) + 2C(i,\,j+1)] - [U(i,\,j) + 2C(i,\,j)] \\
&= \frac{-U(i,\,j)C(i,\,j)}{r(i,\,j)}[t(i,\,j+1) - t(i,\,j)],
\end{aligned}
\quad (4.15)
$$

and

$$
\begin{aligned}
&[U(i,\,j+1) - 2C(i,\,j+1)] - [U(i-1,\,j) - 2C(i-1,\,j)] \\
&= \frac{U(i-1,\,j)C(i-1,\,j)}{r(i-1,\,j)}[t(i,\,j+1) - t(i-1,\,j)].
\end{aligned}
\quad (4.16)
$$

The solution of the above two simultaneous equations yields:

$$
\begin{aligned}
U(i,\,j+1) = &\frac{U(i,\,j+1) + 2C(i,\,j) + U(i-1,\,j) - 2C(i-1,\,j)}{2} \\
&- U(i,\,j)C(i,\,j)\frac{t(i,\,j+1) - t(i,\,j)}{2r(i,\,j)} \\
&+ U(i-1,\,j)C(i-1,\,j)\frac{t(i,\,j+1) - t(i-1,\,j)}{2r(i-1,\,j)}.
\end{aligned}
\quad (4.17)
$$

and

$$C(i, j+1) = \frac{U(i, j) + 2C(i, j) - U(i-1, j) + 2C(i-1, j)}{4}$$

$$- U(i, j)C(i, j)\frac{t(i, j+1) - t(i, j)}{4r(i, j)}$$

$$- U(i-1, j)C(i-1, j)\frac{t(i, j+1) - t(i-1, j)}{4r(i-1, j)}. \tag{4.18}$$

By successive increments of j, points are computed along each characteristic line and the points on each completely computed characteristic are used to compute the next characteristic line given by the next value of i. Thus, the solution is obtained for the entire region of characteristics arising from the "lip" up to the characteristic $\overline{C - D}$, marking the beginning of the third region.

In the third region, as shown in Figure 4.2(b), $\overline{C - D}$ is the characteristic which corresponds to the apex of the lip. This region is called the "expansion fan". This region represents the solution to spreading out due to rarefaction as the leading edge moves inwards onto the dry bed, withdrawing water from the surrounding reservoir. $\overline{M - N}$ in Figure 4.2(a) is the vertical wall. $\overline{C - E}$ and $\overline{C - D}$ are the extreme characteristics which represent the leading edge and the rarefaction edge respectively. The leading edge moves onto the dry bed at a velocity $2\sqrt{g(h + \eta_0)}$ on being released. The variable η_0 here is the elevation of the peak of the lip and is set at h_0. The point C is the fan node. The initial conditions for this region are

$r(i, 1) = R_c \text{(the crater radius)}\,,$

$t(i, 1) = 0\,,$

$C(i, 1) = \sqrt{g(h + \eta^*(i, 1))}\,,$

$\eta^*(i, 1)$ is the elevation of the point on the vertical wall of the crater.

$U(i, 1)$ is specified for each point depending on the location, varying from 0 at the top to $-2\sqrt{g(h + \eta_0)}$ at the bottom. Thus, the maximum slope is for the characteristic from the apex N, along $\overline{C - D}$;

$$\frac{\partial r}{\partial t} = U(i, 1) + C(i, 1)$$

$$= 0 + \sqrt{g(h + \eta_0)}\,, \tag{4.19}$$

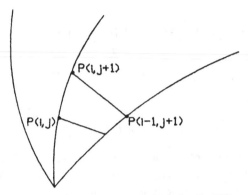

Fig. 4.4. Grid point computation (region III).

and the minimum slope is for the characteristic from the bottom of the crater M, along $\overline{C-E}$:

$$\frac{\partial r}{\partial t} = U(i, 1) + C(i, 1)$$

$$= -2\sqrt{g(h + \eta_0)} + 0. \tag{4.20}$$

The characteristic grid is determined once again except that this time, for the first step, point C is common. The characteristics C+ and C− are discretized with the initial points $P(i, j)$ and $P(i-1, j+1)$ (Figure 4.4) and the solution of the resulting simultaneous equations in r and t gives point $P(i, j+1)$ as

$$t(i, j+1) = \{r(i-1, j+1) - r(i, j) + t(i, j)[U(i, j) + C(i, j)]$$
$$- t(i-1, j+1)[U(i-1, j+1) - C(i-1, j+1)]\}/$$
$$[C(i, j) + U(i, j) - U(i-1, j+1) + C(i-1, j+1)], \tag{4.21}$$

and

$$r(i, j+1) = r(i-1, j+1) + [U(i-1, j+1) - C(i-1, j+1)]$$
$$\cdot [t(i, j+1) - t(i-1, j+1)]. \tag{4.22}$$

Similarly, discretizing Equation (4.6) with $P(i, j)$ and $P(i-1, j+1)$ and solving the resulting equations simultaneously for U and C yields,

$$U(i, j+1) = \frac{U(i, j) + 2C(i, j) + U(i-1, j+1) - 2C(i-1, j+1)}{2}$$

$$- U(i, j)C(i, j)\frac{t(i, j+1) - t(i, j)}{2r(i, j)}$$

$$+ U(i-1, j+1)C(i-1, j+1)\frac{t(i, j+1) - t(i-1, j+1)}{2r(i-1, j+1)}.$$

$$(4.23)$$

and

$$C(i, j+1) = \frac{U(i, j) + 2C(i, j) - U(i-1, j+1) + 2C(i-1, j+1)}{4}$$

$$- U(i, j)C(i, j)\frac{t(i, j+1) - t(i, j)}{4r(i, j)}$$

$$- U(i-1, j+1)C(i-1, j+1)\frac{t(i, j+1) - t(i-1, j+1)}{4r(i-1, j+1)}.$$

$$(4.24)$$

Thus, the complete solution is obtained for the "expansion fan" region.

Finally, the fourth region is ahead of the leading edge of the down side flow from the cavity collapse. Point $P(r, t)$ in this region is situated at a distance from GZ such that the flow has not yet reached that location at time t and the ground is still dry. Thus, the entire collapse has been solved.

To obtain the free-surface elevation or the water-particle velocity at any particular time t^*, a horizontal cut is made in the $(r-t)$ diagram at $t = t^*$ to obtain $P(r, t^*)$. This involves interpolating between the computed U and C at the surrounding mesh node points; once with respect to t and once with respect to r. The surface elevation is given by Equation (4.9). The solution of the cavity collapse for the leading edge holds up to the critical time t_{crit} given by

$$t_{\text{crit}} = \frac{R_c}{2 \cdot C_0}, \qquad (4.25)$$

the time at which the leading edge touches the origin, and the solution becomes unstable due to the singularity in the cylindrical spreading term, $\frac{UC}{r}$, at $r = 0$ in Equations (4.8).

4. Application

As explained in the previous section, the explosion crater may be modeled using a cylindrical cavity with a lip. For the purpose of this test case, a cavity of radius $R_c = 10$ feet, with a lip of length of $R_c/2 = 5$ feet is chosen. Water depth $h = 2.7$ feet and lip height h_0 is to be determined using conservation of mass. The volume of water in the triangular lip on the periphery must equal the water removed from the cylindrical crater.

Volume in the cavity:

$$V_1 = \pi R_c^2 h \,. \tag{4.26}$$

Volume in the lip:

$$V_2 = 2\pi \cdot \left(r + \frac{1}{3}\frac{R_c}{2} \right) \cdot \frac{1}{2} \cdot \frac{R_c}{2} h_0$$

$$= \frac{7}{12}\pi R_c^2 h_0 \tag{4.27}$$

Equating V_1 and V_2 we obtain

$$h_0 = \frac{12}{7}h \,. \tag{4.28}$$

Once again the method of characteristics is applied to the initial condition as described above. Figure 4.5 shows the characteristics grid. Regions I, II, III, and IV, indicating the calm region, the lip evolution, the expansion fan and the dry unaffected regions, are clearly developed. Figure 4.6 shows the free-surface profiles. Evolution of the leading wave from the lip collapse is observed. Care is taken to prevent instability due to the singularity at the origin as r tends to zero.

5. Bore Formation and Propagation

When the leading edge of water strikes GZ, due to the existing axisymmetry, the problem may be analyzed analogously to a supercritical stream of water striking and rebounding off a wall. The process up to the point of contact with the wall is a nondissipative process. But the moment water reaches $r = 0$, turbulence and dissipative reflection on itself begins. The momentum of the incoming flow balances the pressure head of the water accumulating at

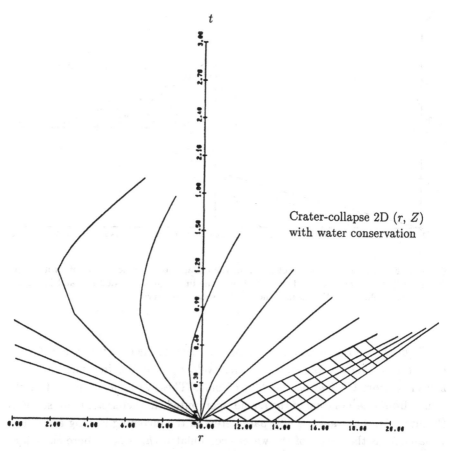

Fig. 4.5. Characteristic grip resulting from the collapse of a cylindrical cavity with a lip. Crater radius is 10 feet, water depth is 2.7 feet. The lip height $h_0 = 4.62$ feet and the lip radius $1 = 5$ feet. To demarcate the regions, only +ve characteristics are plotted in region III.

the center, very similar to a hydraulic jump. However, with continuity being obeyed, the radius of the upper side of the jump increases, seemingly like a translatory bore or a moving hydraulic jump. The average horizontal water particle velocity in the upper side is assumed to be zero. Using the momentum balance approach for the bore formation as described above, the dissipative process is handled externally, without having to analyze the exact internal turbulent flow structure.

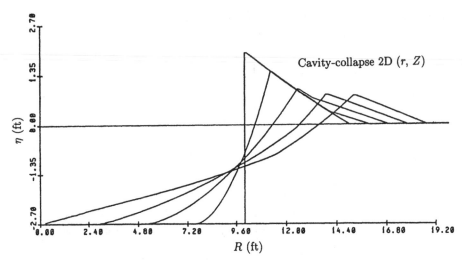

Fig. 4.6. Free-surface profiles resulting from the collapse of a cylindrical cavity with a lip. Crater radius is 10 feet, water depth is 2.7 feet, the lip height $h_0 = 4.62$ feet and lip radius $1 = 5$ feet. Evolution of the lip into the leading wave is observed.

Figure 4.7 shows the schematic representation of the bore formation. Quantities U_d and h_d are the water particle velocity and the free-surface height measured from the bed on the "down" side. These are known quantities obtained from interpolation of the already-computed characteristic mesh solution (Figure 4.5). They are the flow parameters of the incoming leading edge. The variable h_u is the height of the water accumulated ($h_u - h_d =$ bore elevation) and W_b is the bore front velocity of propagation or simply the "bore velocity". The unknowns are h_u and r_u, to be determined as a function of time. W_b is simply $\frac{\partial r_u}{\partial t}$. The governing equations are the continuity and the momentum equation.

Considering Figure 4.8, the subscript 1 denotes known quantities at the beginning of the time step, and subscript 2 indicates the quantities after an interval Δt. The quantity of water flowing into the turbulent region is equal to the increase in volume of the turbulent region, i.e.,

$$h_{u2}\frac{r_2^2}{2} - h_{u1}\frac{r_1^2}{2} = [h_{d2}U_{d2}r_2 \cdot \Delta t + h_{d1}U_{d1}r_1\Delta t]\frac{1}{2}. \qquad (4.29)$$

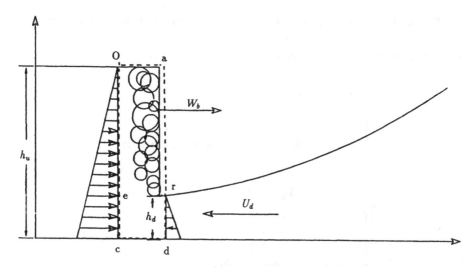

Fig. 4.7. Schematic representation of the turbulent dissipative processes resulting from the rebound of the leading edge of the collapsing crater at the center. "aocd" is the control volume for the application of the momentum theorem. The bore formation is shown with a flat rectangular representation with a frontal velocity W.

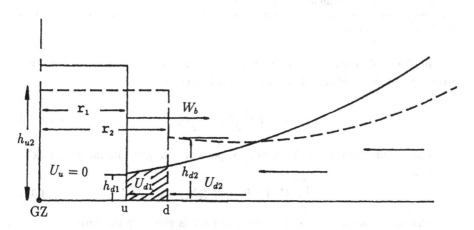

Fig. 4.8. Schematic representation showing the progress of the iterative process over an interval Δt, from 1 to 2. The dashed profile is the advanced solution at 2, and the smooth profile shows the condition at the beginning of the step. The shaded region indicates the unsteady loss term due to the motion of the bore from u to d over the step Δt.

The net inflow momentum per second plus the rate of change of momentum within the control volume due to the translatory term concerning the bore velocity is balanced by the difference of pressure forces.

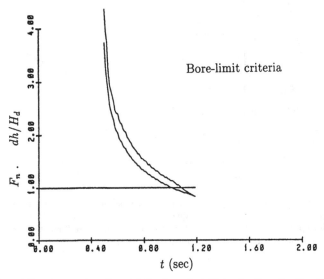

Fig. 4.9. Plot of Froude number Fn and the bore height to depth ratio β, with respect to time. Bore computations are stopped when Fn reaches the value 1. As shown, the condition $Fn < 1$ is reached before $\beta < 0.6$.

For the application of the momentum theorem consider the control volume given by "acod" in Figure 4.9. The external force is

$$\sum Fe = \rho g \left[\frac{1}{2} h_{u2}^2 - \frac{1}{2} h_{d2}^2 \right] . \tag{4.30}$$

A change of momentum with respect to space or change of momentum flux is given as:

$$\rho \iint U \cdot U_n ds = \rho U_{d2}^2 h_{d2} - 0 . \tag{4.31}$$

The change of momentum with respect to time is given by the term,

$$\iiint \frac{\partial(\rho U)}{\partial t} dv = U \frac{\partial M}{\partial t} + M \frac{\partial U}{\partial t} . \tag{4.32}$$

(M and U denote mass of water and velocity.)

The first term $U\frac{\partial M}{\partial t}$ corresponding to the translatory part ("aoer", see Figure 4.7) is zero since $U = 0$ on the upper side. The second term corresponding to the unsteady part "recd" is given by,

$$M\frac{\partial U}{\partial t} = \rho h_{d2} W_b (U_{d2} - 0)\,. \tag{4.33}$$

Equating the external force given by Equation (4.30) with the total change of momentum with respect to time given by Equations (4.31) and (4.33) we obtain,

$$\left[\frac{1}{2}\rho g h_{u2}^2 - \frac{1}{2}\rho g h_{d2}^2\right] = \rho U_{d2}^2 h_{d2} + \rho U_{d2} h_{d2} W_b\,, \tag{4.34}$$

which gives,

$$h_{u2} = \left\{\left[\frac{g h_{d2}^2}{2} + U_{d2}^2 h_{d2} + U_{d2} h_{d2} W_b\right]\frac{2}{g}\right\}^{0.5}\,. \tag{4.35}$$

Equations (4.29) and (4.35) are solved simultaneously using an iterative scheme. Iterative solution of Equation (4.29) gives r_2 the new bore radius. Using the characteristic solution for the down side flow parameters, the new values h_{d2} and U_{d2} are computed by interpolation between the characteristic mesh points. These values are input into Equation (4.35) to obtain h_{u2}. The new value of h_{u2} is refed into Equation (4.29) to recompute r_2. This process is continued until h_{u2} and r_2 converge to a solution.

The solution is advanced in the above manner, step by step. At each step, the Froude Number, $F_n = \frac{U_d}{\sqrt{g h_d}}$, is computed. The bore formation and the process as described above exist only when the Froude Number is less than one. This is used as a criterion to stop the dissipative bore propagation model and to begin wave propagation with a nonlinear nondissipative model. Another criterion for bore existence is the ratio of the bore height to the water depth. Bores are known to disappear when the ratio (Favre, 1935)

$$\beta_b = \frac{h_u - h_d}{h_d} \leq 0.6\,. \tag{4.36}$$

However, our calculations show that the Froude Number criterion is always satisfied first.

Figure 4.9 shows plots of β_b and F_n as a function of time along with the bore-limit criterion to stop the computations. We have the bore height and the radius as a function of time from our previous computation. Free-surface elevations are plotted in Figure 4.10. Time t indicated for each profile is the time elapsed since the release of the cavity. Propagation of the leading wave followed by the bore is shown.

The distance between the leading wave peak and the bore front controls the leading wave length as the bore transforms into an undulatory bore and then into a group of high frequency waves. This is directly dependent on the crater radius. The last wave profile shown in Figure 4.10 is at a time when the flow on the down side just turns subcritical or tranquil. At this time, the wave profile may be used as an input to an appropriate nondissipative wave propagation model.

6. Nonlinear Wave Propagation — Importance of Nonlinearities

It is clear from the observed and recorded wave data close to the explosion in shallow water that the waves exhibit highly peaked wave crests and longer shallow troughs. The leading wave followed by the undular bore, propagates under the influence of the nonlinear convective forces in this region close to GZ. As the wave disperses both angularly and radially, the wave amplitude decays and the existing linear theory is valid only at the limit of the nonlinear range, at a distance from GZ. In this section, the appropriate nonlinear equations are derived in the cylindrical coordinate system, and an improved formula for propagation of transient EGWWs is proposed.

Figure 1.17 shows a typical near-field wave record with the nonlinear characteristics as mentioned before. In order to assess the relative importance of the nonlinearities, one can calculate from the wave records a number of criteria which are grossly related to the ratio of the nonlinear convective inertia to the linear local inertia. One of these is the Ursell parameter

$$U_r = \frac{\eta_0}{L} \left[\frac{L}{h} \right]^3 , \tag{4.37}$$

where η_0 is taken here as the mean leading wave amplitude, L is the linear (long wave) wave length based on the time interval elapsing between the first two crests, and h is the still-water depth. Then, using the wave records obtained from a 10-pound TNT explosion in 4 feet and 2.7 feet of water at a

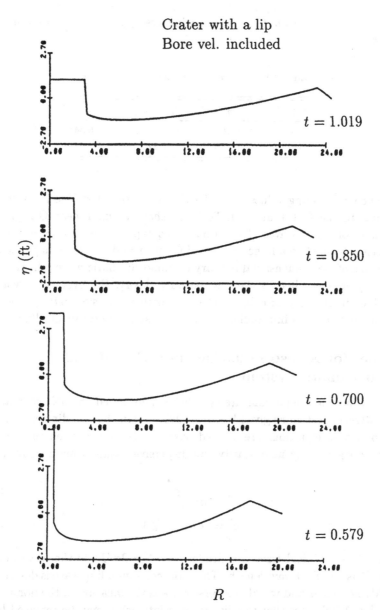

Fig. 4.10. Computed free-surface profiles showing the leading wave followed by a decaying bore. Water depth h is 2.7 feet, the initial crater radius R was 10 feet, lip height h_0 was 4.6 feet, and the lip radius was 5 feet.

distance of 27.7 feet from GZ, the following averaged values are obtained (see Table 4.1.)

Table 4.1. Average leading wave characteristics.

h^* (ft)	η_0^*	T^* (sec)	U_r^*
2.7	0.368	2.053	6.84
4	0.253	2.275	2.63

The corresponding values of the Ursell parameter, $[1 < U < 10$ or $\approx O(1)]$ substantiate the fact that an EGWW in shallow water near GZ enters in the family of water waves of the Boussinesq type. This type of water wave has been subjected to a large number of theoretical developments particularly in the case of two-dimensional Solitary or monochromatic waves. In the latter case, they are described by the Cnoidal wave theory. However, little work has been done in the case of cylindrical waves with radial symmetry. Therefore, Boussinesq equations in a radial coordinate system are established first.

7. The Boussinesq Equations in a Cylindrical Coordinate System

The Boussinesq equations are established here based on their general form in a Cartesian system as given by Mei (1983). As in the two-dimensional case, the Boussinesq equations are derived based on the assumption that the two parameters governing nonlinearity and dispersive nature remain small, i.e.,

$$\varepsilon = \frac{A_\eta^*}{h^*} \ll 1 \tag{4.38}$$

$$\mu^2 = (k^* h^*)^2 \ll 1 \tag{4.39}$$

where A_η^* is a typical wave amplitude and $k^* = 2\pi/L^*$ is the wave number, where L^* is a typical wave length. Then, the corresponding wave is described as "weakly nonlinear and weakly dispersive". ε is the measure of the nonlinearity and $\mu^2 \ll 1$ indicates that the dispersion relationship may be replaced by the long-wave approximation. The above assumption that $O(\varepsilon) = O(\mu^2) \ll 1$ implies that the Ursell parameter

$$U_r = \frac{\varepsilon}{\mu^2} = O(1). \tag{4.40}$$

In order to establish Boussinesq equations in the general cartesian (x, y, z, t) system, all the quantities are nondimensionalized as follows (the asterix refers to dimensional quantities):

$$(x, y) = k^*(x^*, y^*) \qquad \left(k^* = \frac{2\pi}{L^*}\right). \tag{4.41}$$

The dimensionless time is

$$t = t^*(gh^*)^{\frac{1}{2}}k^*, \tag{4.42}$$

where g is the gravity acceleration. The dimensionless free-surface elevation is,

$$\eta = \frac{\eta^*}{A^*}, \tag{4.43}$$

where A^* is the amplitude. The dimensionless potential function is given by

$$\phi = \phi^* \left[\frac{A^*}{k^*h^*}\sqrt{gh^*}\right]^{-1}. \tag{4.44}$$

The horizontal and vertical velocity components are expressed as

$$\mathbf{U}(u, v) = \frac{h^*}{A^*\sqrt{gh^*}}\mathbf{U}(u^*, v^*) \tag{4.45}$$

and

$$w = \frac{k^*h^*}{A^*}\frac{h^*}{\sqrt{gh^*}}w^*. \tag{4.46}$$

Here, u^*, v^*, and w^* are the velocity components in the x, y, and z directions respectively. The solution satisfies continuity and, therefore, under the assumption of irrotationality, the Laplace equation

$$\mu^2(\phi_{xx} + \phi_{yy}) + \phi_{zz} = 0 \tag{4.47}$$

is obeyed for $-1 < z < \varepsilon\eta$ and ∇ is defined to operate only the horizontal plane

$$\nabla = i\frac{\partial}{\partial x} + i\frac{\partial}{\partial y}. \tag{4.48}$$

The kinematic boundary condition at the free surface ($z = \varepsilon\eta$) is

$$\mu^2(\eta_t + \varepsilon\phi_x\eta_x + \varepsilon\phi_y\eta_y) = \phi_z\,. \tag{4.49}$$

The dynamic free-surface condition (at $z = \varepsilon\eta$) is

$$\mu^2(\phi_t + \eta) + \frac{1}{2}\varepsilon[\mu^2(\phi_x^2 + \phi_y^2) + \phi_z^2] = 0\,. \tag{4.50}$$

The impermeable bottom condition is written as

$$\phi_z = 0 \quad \text{at} \quad z = -1\,. \tag{4.51}$$

Combining Equation (4.48) with Equations (4.49), (4.50), and (4.51) using a series solution for ϕ and neglecting all the terms of $O(\varepsilon^2,\ \mu^2$ and above), the following Boussinesq equations are obtained:

$$\eta_t + \nabla \cdot [(\varepsilon\eta + 1)\mathbf{U}] = 0\,, \tag{4.52}$$

$$\mathbf{U}_t + \varepsilon\mathbf{U}\cdot\nabla\mathbf{U} + \nabla\eta - \frac{\mu^2}{3}\nabla(\nabla \cdot \mathbf{U}_t) = 0\,. \tag{4.53}$$

Replacing $\mathbf{U}$ by $\nabla\phi$ and taking the derivative of Equation (4.53) with respect to t and submitting Equation (4.52) to a ∇ operator, η can be eliminated from Equation (4.52) so that

$$\phi_{tt} + \frac{\varepsilon}{2}(\nabla\phi \cdot \nabla\phi)_t + \varepsilon\nabla \cdot (\phi_t\nabla\phi) - \nabla^2\phi = \frac{\mu^2}{3}[\nabla \cdot \nabla\phi_{tt}]\,, \tag{4.54}$$

which is the alternate form of the Boussinesq equation.

For the problem under consideration, the equations derived above are easily transformed into a cylindrical coordinate system (r, ϕ, z) with axisymmetry. Considering the fact that partial derivatives with respect to θ are zero, the following relations may be written:

$$\nabla\phi = i\phi_r \tag{4.55}$$

$$\nabla \cdot \phi = \frac{1}{r}(r\phi)_r \tag{4.56}$$

$$\nabla^2\phi = \frac{1}{r}(r\phi_r)_r\,. \tag{4.57}$$

Inserting Equations (4.55)–(4.57) into (4.52) and (4.53), the Boussinesq equations in cylindrical coordinates are obtained as

$$\eta_t = \frac{1}{r}[r(\varepsilon\eta + 1)\phi_r]_r = 0, \tag{4.58}$$

$$U_t + \varepsilon U \cdot U_r + \eta_r = \frac{\mu^2}{3}\left[\frac{1}{r}(r \cdot \phi_r)_r\right]_{rt}. \tag{4.59}$$

These equations are identical to those obtained by Chwang and Wu (1976). They are also equivalent to the one established by Le Méhauté *et al.* (1987), with some simple transformations. Eliminating η, as previously done in the Cartesian system, yields the alternate Boussinesq equation in a cylindrical coordinate system

$$\phi_{tt} - \phi_{rr} - \frac{1}{r}\phi_r + \varepsilon(2\phi_r\phi_{rt} + \frac{1}{r}\phi_t\phi_r + \phi_t\phi_{rr}) = \frac{\mu^2}{3}\left[\phi_{rrtt} + \frac{1}{r}\phi_{rtt}\right]. \tag{4.60}$$

Given η and using the relationship

$$\eta = -\phi_t + O(\varepsilon, \mu^2), \tag{4.61}$$

Equation (4.60) can be solved numerically. However, this task is better accomplished by using a moving coordinate system traveling at the phase speed. The Boussinesq equations are then transformed into the KdV (Korteweg and de Vries) equation as follows.

8. Transformation of the Boussinesq Equations into KdV Equations

A moving coordinate system in σ and τ is defined using the relations,

$$\begin{aligned}\sigma &= r - t \\ \tau &= \varepsilon t,\end{aligned} \tag{4.62}$$

so that the phase is given by

$$\xi = r - t + O(\varepsilon). \tag{4.63}$$

Thus, the observer moving at $C = \frac{C^*}{\sqrt{gh^*}}$ observes only a slow variation in the wave form with respect to time. So, we have the forward transform operator pair to transform the governing equations from the stationary coordinate system to a moving coordinate system $[(r, t) \to (\sigma, \tau)]$,

$$\frac{\partial}{\partial r} = \frac{\partial}{\partial \sigma} \,,$$

$$\frac{\partial}{\partial t} = \varepsilon \frac{\partial}{\partial \tau} - \frac{\partial}{\partial \sigma} \,,$$

$$(4.64)$$

and the reverse transform pair to go from the moving coordinate system back to the stationary coordinate system

$$\frac{\partial}{\partial \sigma} = \frac{\partial}{\partial r} \,,$$

$$\frac{\partial}{\partial t} = \frac{1}{\varepsilon} \frac{\partial}{\partial t} + \frac{1}{\varepsilon} \frac{\partial}{\partial r} \,.$$

$$(4.65)$$

Inserting these forward transform operators into Equation (4.60) and retaining the terms only of $O(\varepsilon, \mu^2)$, one obtains,

$$\phi_{\sigma r} + \frac{\phi_\sigma}{2\varepsilon r} + \frac{3}{4}(\phi_\sigma^2)_\sigma + \frac{(\phi_\sigma)^2}{2r} + \frac{\mu^2}{6\varepsilon} \left[\phi_{\sigma\sigma\sigma\sigma} + \frac{\phi_{\sigma\sigma\sigma}}{r} \right] = 0 \,. \qquad (4.66)$$

Also,

$$\phi_\sigma = -\phi_t = \eta + O(\varepsilon, \mu^2) \qquad (4.67)$$

from the dynamic boundary condition. Therefore, ϕ_σ can be replaced by η at the leading order of approximation. Thus, Equation (4.66) becomes

$$\eta_r + \frac{\eta}{2\varepsilon r} + \frac{3}{2}\eta\eta_\sigma + \frac{\eta^2}{2r} + \frac{\mu^2}{6\varepsilon} \left[\eta_{\sigma\sigma\sigma} + \frac{\eta_{\sigma\sigma}}{r} \right] = 0 \,. \qquad (4.68)$$

Reverting back to the stationary coordinates using Equation (4.65) one obtains

$$\eta_t + \left[1 + \frac{3}{2}\varepsilon\eta \right] + \left[\frac{\eta}{2r} + \frac{\eta^2 \varepsilon}{2r} + \frac{\mu^2}{6} \frac{\eta_{rr}}{r} \right] + \frac{\mu^2}{6}\eta_{rrr} = 0 \,. \qquad (4.69)$$

Equation (4.69) leads to two forms of the KdV equation. A stretched system of coordinates given by

$$r' = \varepsilon r$$

$$t' = t - r$$

$$(4.70)$$

is introduced such that the operator pair transforming to the stretched coordinates is

$$\frac{\partial}{\partial r} = t\frac{\partial}{\partial r'} - \frac{\partial}{\partial t'}$$ (4.71)

$$\frac{\partial}{\partial t} = \frac{\partial}{\partial t'}.$$ (4.72)

Substituting these into Equation (4.69) and dropping all the terms of order higher than ε and μ^2, one obtains

$$\frac{\partial \eta}{\partial r'} - \frac{3}{2}\eta\frac{\partial \eta}{\partial t'} + \frac{\eta}{2r'} - \frac{\mu^2}{6\varepsilon}\frac{\partial^3 \eta}{\partial t'^3} = 0.$$ (4.73)

Reverting back to the normal coordinates finally gives us the first form of the KdV equation:

$$\frac{\partial \eta}{\partial r} + \left(1 - \frac{3}{2}\varepsilon\eta\right)\frac{\partial \eta}{\partial t} + \frac{\eta}{2r} - \frac{\mu^2}{6}\frac{\partial^3 \eta}{\partial t^3} = 0.$$ (4.74)

It is possible to arrive at another equivalent form of the KdV equation by introducing once again the σ and τ coordinate system, Equation (4.62) into Equation (4.74) and keeping terms of only $O(\varepsilon, \mu^2)$ to get

$$\eta_r + \frac{3\eta\eta_\sigma}{2} + \frac{\eta}{2r\varepsilon} + \frac{\mu^2}{6\varepsilon}\eta_{\sigma\sigma\sigma} = 0.$$ (4.75)

Reverting back to stationary coordinates yields the second form of the KdV equation:

$$\eta_t + \left(1 + \frac{3}{2}\varepsilon\eta\right) + \frac{\eta}{2r} + \frac{\mu^2}{6}\eta_{rrr} = 0.$$ (4.76)

Equations (4.74) and (4.76) are two equivalent forms the KdV equation in the nondimensional coordinates systems. Rewriting them in terms of dimensional quantities, two canonical forms of KdV equations in cylindrical coordinates are obtained. These are

$$\frac{\partial \eta^*}{\partial t^*} + \sqrt{gh^*}\left(1 + \frac{3\eta^*}{2h^*}\right)\frac{\partial \eta^*}{\partial r^*} + \frac{\eta^*}{2r^*}\sqrt{gh^*} + \sqrt{gh^*}\frac{h^{*2}}{6}\frac{\partial^3 \eta^*}{\partial r^{*3}} = 0,$$ (4.77)

$$\frac{\partial \eta^*}{\partial r^*} + \left(1 - \frac{3\eta^*}{2h^*}\right)\frac{1}{\sqrt{gh^*}}\frac{\partial \eta^*}{\partial t^*} + \frac{\eta^*}{2r^*} - \frac{h^{*2}}{6(gh^*)^{\frac{3}{2}}}\frac{\partial^3 \eta^*}{\partial t^{*3}} = 0.$$ (4.78)

9. Extended KdV Equation

The validity of the previously established KdV equation depends on the relative importance of the nonlinear convective terms which are grossly characterized by the Ursell parameter. Based on this, one has seen that the KdV equation seems to be most appropriate to describe the leading waves of EGWWs in shallow water. However, the theory is seldom valid for trailing waves in deeper water which are better approximated by the Stokesian theory. In order to prevent the use of two separate families of wave theories and keeping the same validity for the leading waves as well as trailing waves in shallow or intermediate water depth, a modification to the classical KdV equation is introduced through the dispersion relationship.

To analyze the phase speed of wave propagation, we consider the linearized KdV equations, without the cylindrical spreading for the sake of simplicity. Equation (4.77) becomes

$$\frac{\partial \eta^*}{\partial t^*} + \sqrt{gh^*}\frac{\partial \eta^*}{\partial r^*} + \frac{h^{*2}}{6}\sqrt{gh^*}\frac{\partial^3 \eta^*}{\partial r^{*3}} = 0 \,, \tag{4.79}$$

and Equation (4.78) can be written as

$$\frac{\partial \eta^*}{\partial r^*} + \frac{1}{\sqrt{gh^*}}\frac{\partial \eta^*}{\partial t^*} - \frac{h^{*2}}{6}\frac{1}{(gh^*)^{\frac{3}{2}}}\frac{\partial^3 \eta^*}{\partial t^{*3}} = 0 \,. \tag{4.80}$$

The above equations yield the dispersion relationships

$$\omega^* = \sqrt{gh^{*2}}\left(k^* - \frac{h^{*2}}{6}k^{*3}\right), \tag{4.81}$$

and

$$k^* = \frac{\omega^*}{\sqrt{gh^*}} + \frac{h^{*2}}{6(\sqrt{gh^*})^3}\omega^3 \,, \tag{4.82}$$

respectively. Figure 4.11 is based on the dispersion relations (4.81) and (4.82) along with linear wave dispersion given by

$$\omega^{*2} = gk^* \tanh(k^*h^*) \,. \tag{4.83}$$

Clearly, it is seen that the linear dispersion relationships given by Equations (4.81) and (4.82) match the classical Airy theory equation (4.83) for small values of k^*h^*, but they lose their validity for the larger values. In order to

DISPERSION RELATION

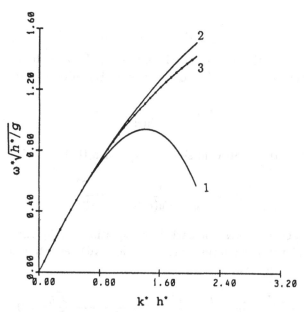

Fig. 4.11. A comparison of the nondimensional wave frequency as a function of nondi-mensional wave number as given by the three dispersion relations. 1, 2, and 3 refer to Equations (4.84), (4.85), and (4.86) respectively.

expand the validity of the KdV equation into intermediate water depth, an extended form of the KdV equation is presented. This incorporates the linear wave dispersion relationship given by the Airy theory into the KdV equations instead of the long-wave approximation. This allows a smooth transition from shallow water nonlinear waves to intermediate and deep water waves. The nonlinearity is a function of wave height and is also expected to vanish smoothly as the wave amplitudes decay with distance. The extended KdV equation was first proposed by Whitham (1974) for water waves using integro differential equations. In the context of the split step method described in the following section, Tappert and Judice (1972) wrote the extended KdV equation for an ion acoustic wave. An approach based on the above two works is followed here. The linear equation (4.80) may be written as:

$$\frac{\partial \eta^*}{\partial r^*} + i\Omega_{op}\eta^* = 0 \,. \tag{4.84}$$

Here,

$$i\Omega_{op} = \frac{1}{\sqrt{gh^*}}\frac{\partial}{\partial t^*} - \frac{h^{*2}}{6(gh^*)^{3/2}}\frac{\partial^3}{\partial t^{*3}} \,, \tag{4.85}$$

$\Omega_{\sigma\rho}$ corresponds to the dispersion relation (4.82). In order to obtain the operator for the full wave dispersion, one can write Equation (4.83)

$$k^* = \frac{1}{g \cdot \tanh(k^* h^*)}\omega^2 \,. \tag{4.86}$$

ω is replaced on the right-hand side by $-i\frac{\partial}{\partial t}$ such that

$$i\Omega_{op} = \frac{1}{g \cdot \tanh(k^* h^*)}(-i)\frac{\partial^2}{\partial t^{*2}} \,. \tag{4.87}$$

Putting the above expression back into Equation (4.85) and including the spreading and nonlinear terms, the Extended KdV equation for time series analysis is obtained. Thus,

$$\frac{\partial \eta^*}{\partial r^*} - \frac{3\eta^*}{2h^*}\frac{1}{\sqrt{gh^*}}\frac{\partial \eta^*}{\partial t^*} + \frac{\eta^*}{2r^*} - \frac{i}{g\,\tanh(k^* h^*)}\frac{\partial^2 \eta^*}{\partial t^{*2}} = 0 \,. \tag{4.88}$$

10. Modes of Solutions and Numerical Procedures — The Split-Step Fourier Algorithm

Analytical two-dimensional steady state solutions of the KdV equation have already been obtained. These are the Conoidal wave and Solitary wave solutions (Whitham, 1974). However, in the case of transient water waves in three-dimensions, with the presence of a cylindrical spreading term, analytical solutions are not possible. Numerical solutions have been developed and tested with experimental data by many investigators. The following numerical methods can be used:

(1) Finite Difference Technique,
(2) Numerical Integration,
(3) Split-Step Fourier Algorithm (SSFA).

Peregrine (1967) gave finite difference schemes to solve both KdV and Boussinesq equations in two dimensions. Chwang and Wu (1976) used a similar scheme to solve the Boussinesq equations in a cylindrical coordinate system. Solutions to the cylindrical KdV equations, first derived by Miles (1978) in the context of free-surface gravity waves, was experimentally tested by Weidman and Zakhem (1988). Numerical solutions for the same cylindrical KdV equation was given earlier by Ko and Kuehl (1979) using numerical integration. The SSFA has not found as much application in free-surface gravity waves as in plasma waves. The only application was by Chang *et al.* (1979), who used the SSFA to solve the KdV equation over a slowly varying channel. They found that the SSFA gave solutions with a high degree of accuracy. All the aforementioned solutions made use of a solitary wave as the input for analysis. The experimental verification was also carried out for solitary waves. The numerical scheme shows a high level of accuracy for the propagation of solitary waves.

For the problem of explosion-generated water waves, propagation over large distances and many wave periods is expected. Typical values for Δt and Δx used in the finite difference schemes were $\Delta t = \Delta x \approx 0.2$. Limitations imposed by stability and convergence criteria make both finite difference and numerical integration schemes inadequate. Therefore, the SSFA method is adopted to propagate real wave records in cylindrical coordinates for the first time in view of its particularly suitable computational scheme for EGWWs.

Explosion test data recorded are in the time series of the free-surface elevation recorded at various distances from GZ. Thus, Equation (4.78) which is suitable for time series analysis is retained. It may also be reorganized as

$$\frac{\partial \eta}{\partial r} - \frac{3}{2} \frac{\eta}{h} \frac{1}{\sqrt{gh}} \frac{\partial \eta}{\partial t} + \frac{\partial \eta}{\partial t} \frac{1}{\sqrt{gh}} + \frac{\eta}{2r} - \frac{h^2}{6(gh)^{3/2}} \frac{\partial^3 \eta}{\partial t^3} = 0 \qquad (4.89)$$

("$*$" indicates that dimensional quantities are removed for convenience). The essence of SSFA lies in separating Equation (4.89) in two parts,

$$\frac{\partial \eta}{\partial r} - \frac{3}{2} \frac{\eta}{h} \frac{1}{\sqrt{gh}} \frac{\partial \eta}{\partial t} = 0, \qquad (4.90)$$

and

$$\frac{\partial \eta}{\partial r} + \frac{\eta}{2r} + \frac{\partial \eta}{\partial t} \frac{1}{\sqrt{gh}} - \frac{h^2}{6(gh)^{3/2}} \frac{\partial^3 \eta}{\partial t^3} = 0. \qquad (4.91)$$

Equation (4.90) consists of only the nonlinear part which is perceived as the nonlinear correction part of the KdV equation. Equation (4.91) is the linear dispersive part. In the absence of nonlinear terms, solutions represented by Equation (4.91) follow the long wave approximation in shallow water. The solution is obtained in two steps. Step I involves solution of Equation (4.90) using an implicit iterative finite difference scheme. The solution is advanced by a step Δr, $(r \rightarrow r + \Delta r)$, using Equation (4.90). In Step II, the corrected time series is advanced using Equation (4.91) over the same step in r. Thus, Steps I and II, which yield partial solutions, are combined by performing individual operations in sequence to give the complete solution.

Step I: Nonlinear Correction

The solution of the nonlinear equation (4.90) is done in two steps in order to set up the implicit finite difference scheme.

Step Ia

For convenience we use the following notation:

$$\eta(r,\, t) = \eta_i^j\,,$$

$$i \equiv r\,,$$

$$j \equiv t\,,$$

the solution is advanced in r using simple forward deference. Equation (4.90) is split as

$$\frac{\eta_{i+1}^j - \eta_i^j}{\Delta r} = \frac{\eta_i^{j+1} - \eta_i^{j-1}}{2\Delta t \cdot \sqrt{gh}} \cdot \frac{3}{2h} \cdot \eta_i^j\,. \tag{4.92}$$

This gives

$$\eta_i^j + 1 = \eta_i^j + \frac{\Delta r}{\Delta t} \cdot \frac{3}{4h\sqrt{gh}}(\eta_i^{j+1} - \eta_i^{j-1})\eta_i^j\,. \tag{4.93}$$

Step Ib

The availability of $\eta(r + \Delta r)$ now enables us to split Equation (4.90) using an implicit finite difference scheme as follows:

$$\frac{\eta_{i+1}^j - \eta_i^j}{\Delta r} = \frac{(\eta_{i+1}^{j+1} - \eta_{i+1}^{j-1} + \eta_i^{j+1} - \eta_i^{j-1})}{4\Delta t \sqrt{gh}} \frac{3}{2h} \cdot \eta_i^j\,, \tag{4.94}$$

which gives

$$\eta_{i+1}^{*j} = \eta_i^j + \frac{\Delta r}{\Delta t} \cdot \frac{3}{8h\sqrt{gh}} (\eta_{i+1}^{j+1} - \eta_{i+1}^{j-1} + \eta_i^{j+1} - \eta_i^{j-1})\eta_i^j \,. \tag{4.95}$$

η_{i+1}^j is obtained from Equation (4.93) and is fed into the right-hand side of Equation (4.95) to obtain η_{i+1}^{*j}. This is once again fed into the right-hand side of Equation (4.95) in an iterative loop until a sufficient level of accuracy has been achieved. The time series η_{i+1}^{*j} is the corrected time series.

Step II: Linear Propagation

In this step, the corrected time series is propagated using the Fourier transform solution of the linear partial differential equation (4.91). We define the Fourier transform pair as

$$\bar{\eta}(\omega) = \int_{-\infty}^{\infty} \eta(t)e^{i\omega t}dt \qquad FFT^+ \tag{4.96}$$

and

$$\eta(t) = \frac{1}{2\pi} \int_{-\infty}^{\infty} \bar{\eta}(\omega)e^{-i\omega t}d\omega \qquad FFT^- \,. \tag{4.97}$$

A forward Fourier transform is applied in Equation (4.91) so that

$$\frac{\partial \bar{\eta}}{\partial r} + \frac{\bar{\eta}}{2r} + \frac{(i\omega)}{\sqrt{gh}}\bar{\eta} - (-i\omega)^3 \cdot \alpha\bar{\eta} = 0 \,, \tag{4.98}$$

where

$$\alpha = \frac{h^2}{6(\sqrt{gh})^3} \,. \tag{4.99}$$

Separating η to one side, we obtain the first order ordinary differential equation in $\bar{\eta}$ and r:

$$\frac{\partial \bar{\eta}}{\bar{\eta}} = \left[-\frac{1}{2r} + \frac{i\omega}{\sqrt{gh}} + i\alpha\omega^3 \right] dr \,. \tag{4.100}$$

Integrating the above equation from r to $r + \Delta r$, we obtain

$$\bar{\eta}(r + \Delta r, \, \omega) = \sqrt{\frac{r}{r + \Delta r}} e^{i(\frac{\omega\Delta r}{\sqrt{gh}} + \alpha\omega^3 \Delta r)}\bar{\eta}(r, \, \omega) \,. \tag{4.101}$$

The inverse Fourier transform on both sides yields

$$\eta(r + \Delta r, t) = \sqrt{\frac{r}{r + \Delta r}} \, FFT^{-1} \left[\bar{\eta}(r, \omega) e^{i\left(\frac{\omega \Delta r}{\sqrt{gh}} + \alpha \omega^3 \Delta r\right)} \right], \qquad (4.102)$$

which is the same as

$$\eta^j_{i+1} = \sqrt{\frac{r}{r + \Delta r}} \, FFT^{-1} \left\{ [FFT^{+1}(\eta^{*j}_{i+1})] e^{i\left(\frac{\omega \Delta r}{\sqrt{gh}} + \alpha \omega^3 \Delta r\right)} \right\}. \qquad (4.103)$$

Note that $\bar{\eta}(r, \omega)$ on the right-hand side of Equation (4.102) has been replaced by the Fourier transform of the corrected time series from Step I. η^j_{i+1} from Equation (4.103) is the final advanced solution.

11. Theoretical and Experimental Verification

Two computer codes have been developed based on Equations (4.77) and (4.78) respectively, using the SSFA as described in the previous section. The first one, "KdV.FOR" is based on Equation (4.78) in which case a time series of the free-surface elevation at a fixed distance r can be used as input. The output is then the time series of the free-surface elevation $\eta(t)$ at a given distance r from GZ. The second one, "KdVR.FOR", is based on Equation (4.77). The input is the free-surface elevation at time $t = 0$ given as a function of r. It generates the free-surface elevation, $\eta(r)$, at any time $t > 0$.

In order to verify the "KdVR.FOR" program, the evolution of a solitary wave in a wedge was investigated so as to be able to duplicate the results obtained by other methods (Chwang and Wu, 1976). The peak amplitude and the phase speed match the values obtained by Chwang and Wu very well.

The "KdV.FOR" program is also verified by duplicating the results of Weidman and Zakhem (1988) where the input is the time history of the free-surface elevation. For outgoing waves, a dispersive tail appears. The theoretical law of decay of wave height $(r^{-2/3})$ is also well verified. Therefore, both models are verified and substantiate the results previously obtained for cylindrical solitary waves. The computer codes are both accurate and efficient.

The previous theory has been applied to a series of wave records obtained from underwater explosions. The experiments took place in a shallow water pond at the Coastal Engineering Research Center, Waterways Experiment Station, Vicksburg, Mississippi. The basin is rectangular (100 feet × 140 feet) and surrounded by a levee with a 3/2 slope. A series of tests were performed, varying the yield, the depth of burst, and the water depth. The explosions took place on a thick 30 feet by 30 feet concrete horizontal slab so as to prevent the projection of mud which would add significant noise to the wave records. Using the models constructed above, it is possible to propagate waves generated by an explosion in the near-field. To test the validity of the numerical scheme,

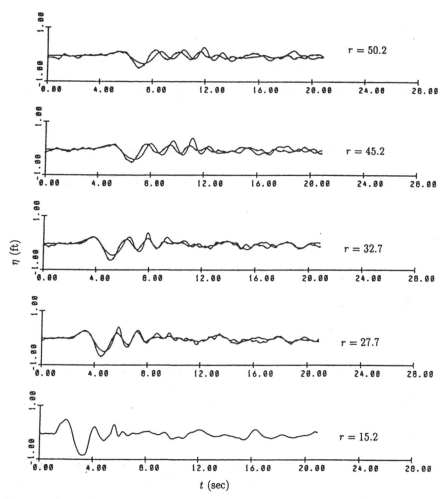

Fig. 4.12. Comparison of experimental wave records with the theory using the wave record at $r = 15.2$ feet as input. The yield $W = 10$ pounds, water depth $h = 2.7$ feet, and the depth of burst $z = 0.6$ feet.

the first time series at $r = 15.2$ feet is used as the input. Then, at subsequent intervals, the experimental wave record is compared with the one generated numerically. Figure 4.12 is an example of these comparisons. The numerical solutions seem to match the experimental data very well, reproducing the nonlinearities in terms of steep crests and shallow troughs. However, a small

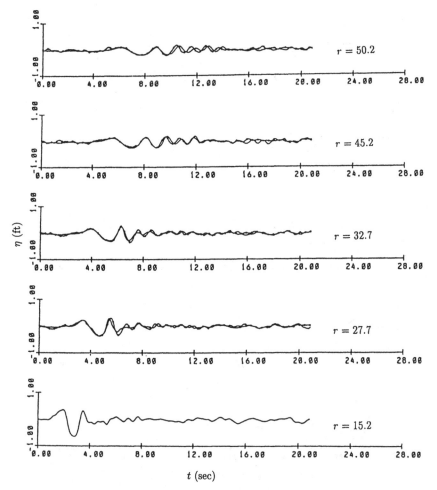

Fig. 4.13. Comparison of experimental wave records with the theory using the wave record at $r = 15.2$ feet as input. The yield $W = 10$ pounds, water depth $h = 2.7$ feet, and the depth of burst $z = 0.6$ feet. The adjusted water depth is 2.2 feet.

phase difference between the numerical and experimental solution is evident, indicating that there is a small difference in the phase speed of propagation. This has been observed in other cases. This could be accounted for by the fact that experiments were done on a muddy bottom which could damp the phase speed of propagation. Another reason could be faulty reading of water depth. By adjusting the water depth, it was possible to externally reduce the phase speed of the numerical waves to that of experimental waves to obtain a good match between theory and data as shown in Figure 4.13. Therefore, the question can be raised about the accuracy of the depth measurements. The water depth was measured at GZ where a thick 30 feet by 30 feet horizontal concrete slab was built as previously mentioned. Beyond this slab, the rest of the basin is silt and fine sand, which could have been moved by successive explosions. Therefore, the water depth may vary from one area to another. This affects not only the phase velocity but it may also induce some wave dissymmetry by refraction. If one adds the possibility of the effect of wave-bottom interaction resulting from the wave-induced pressure fluctuation and boundary layer, it is concluded that the theory is verified by experiments within the accuracy of the experimental errors.

For further confirmation of the theory, another series of tests were conducted entirely on a concrete slab to minimize the previously described errors. However, before presenting these results, it is interesting to determine the relative importance of the nonlinear terms. This is brought about by running the computer program without the nonlinear terms. Even after depth adjustments, the numerical solution is unable to match the data when the nonlinear terms are suppressed. Although nonlinear time series are used as input, the nonlinearity soon disappears and the wave form adjusts itself symmetrically about the mean water level. Figure 4.14 shows the comparison.

Nonlinear waves in shallow water were also generated by dropping a circular plate in very shallow water to simulate waves like those produced in an explosion. These were done over a concrete bottom, thus water depth measurements may be assumed to be accurate.

Once again the time series, both experimental and numerically generated using the KdV solver, were compared. This time the two matched perfectly.

It is possible to run the models in the reverse direction. That is a far-field wave record can be used as input to generate a time series at near field. Figure 4.15 shows a comparison between the experimental data and numerical solution, using far-field data as input and running the model in the reverse

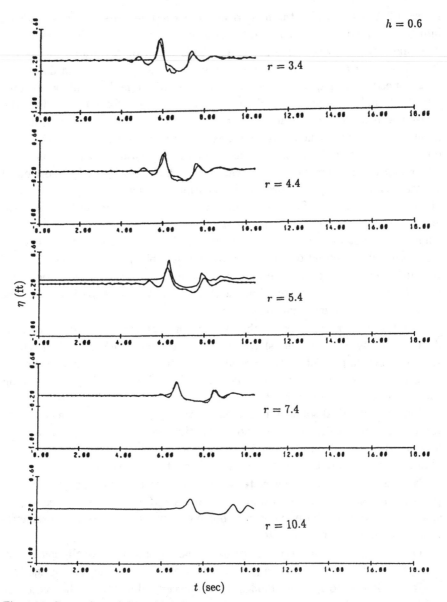

Fig. 4.14. Comparison of the results obtained from the KdV equation without the nonlinear terms. The dotted line is the linear theory.

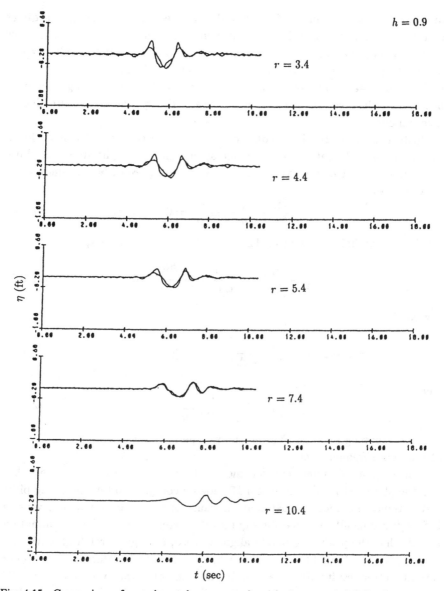

Fig. 4.15. Comparison of experimental wave records with the extended KdV theory using the wave records at $r = 10.4$ as input and running the model backwards (negative time steps). Steel plate diameter $D_s = 2.8$ feet, thickness 1 inch, water depth $h = 0.9$ feet and the drop height $zl = 3$ feet.

direction. Judging from the fact that the KdV equation solver gave a direct match between the data and numerical solution for the flat plate, this indicates that the KdV equation is a good model for propagating nonlinear shallow water waves. It also shows that the depth adjustments, which are necessary for 10-pound shots, were to account for external effects like soil damping or uneven bottom. Differences in phase speed between theory and data were due to external effects.

Instead of using the KdV equation, we now use the extended KdV form, such as given in Equation (4.88). Once again, using the split step procedure, the nonlinear part of the extended KdV Equation (4.87) ("*" is removed for convenience),

$$\frac{\partial \eta}{\partial r} - \frac{3\eta}{2h} \frac{1}{\sqrt{gh}} \frac{\partial \eta}{\partial t} = 0 , \tag{4.104}$$

is solved using finite difference. The solution of the linear part,

$$\frac{\partial \eta}{\partial r} + \frac{\eta}{2r} - \frac{i}{g \cdot \tanh(kh)} \frac{\partial^2 \eta}{\partial t^2} = 0 \tag{4.105}$$

is given by

$$\eta(r + \Delta, t) = \sqrt{\frac{r}{r + \Delta r}} FFT^{-1}[\bar{\eta}(r, \omega)e^{-i\alpha\omega^2 \Delta r}] , \tag{4.106}$$

where

$$\alpha = \frac{1}{g \tanh[k(\omega)h]} . \tag{4.107}$$

Split stepping and the Fourier analysis permits the iterative computation of k for each frequency ω.

Once again this numerical scheme has been tested with a 10-pound shot and the plate test results. Figure 4.15 shows a comparison for the falling plate test without any depth adjustment. Because of the good match that we had before, it is difficult to notice any significant difference. But a close examination reveals that the peaks are better aligned. Also, a longer stretch of waves are better matched than before. So as waves move into intermediate or deeper waters, better results are expected. It must be noted that the extended KdV equation used here is an approximation. But since the KdV equation itself is an approximation, its extension is justified on the basis that the solution is improved and its range expanded.

Chapter 5

APPLICATION OF NONLINEAR
THEORY AND CALIBRATION

1. Evolution from Bore to Nonlinear Waves

Chapter 4 demonstrated the ability of the cylindrical KdV equation to predict the propagation of real EGWWs. The wave generation phenomenon discussed in Chapter 4 was carried to a point with the intention that a suitable nonlinear propagation theory would be applied thereafter to generate and propagate waves. In this chapter, the wave generation mechanism is continued using the KdV solvers. The input to the KdV equation are the wave records that resulted from the cavity collapse and the resulting bore formation.

Following the wave evolution process, an attempt to calibrate the initial condition, namely the crater radius R, is made. This is done on the basis that the leading wave period or the distance between the leading wave and the second wave is directly proportional to the crater size. Therefore, by adjusting the crater radius, the leading wave characteristics are matched with the recorded real EGWWs. This is presented as a preliminary estimate of the physical crater resulting from the explosion. The crater radius is calibrated with the yield of the explosion. Hydrodynamic dissipation is computed and efficiency of underwater explosions in shallow water is given.

193

Based on the method of characteristics and "bore" computations, water elevation is presented in Figure 4.10. The computations are stopped at $t = 1.019$, based on the "bore limit" criterion. Now this wave record as a function of r is to be used as the input (initial condition) to propagate the resulting waves using the KdV equations. The appropriate KdV equation is Equation (4.78) which allows computation of wave elevation after each successive time interval. Figure 5.1 shows the wave record from bore computations evolving into undulatory waves. The leading wave loses its sharp peak, and the rectangular representation of the bore takes on a favorable undulatory wave-like character similar to a translatory undular nondissipative bore.

In view of the fact that available real explosion wave records are in the form of time series, one must convert from a wave elevation distribution in r at a particular time t (wave snap-shot), to a time history of wave elevation at a particular distance r from GZ. This is achieved quite simply, owing to the numerical computation scheme of solving the KdV equation. Referring to Figure 5.1, in computation of a wave record at $t = 1.019$ to the one at $t = 11.019$ using Equation (4.78), a vertical cut is made at a specified r ($r_0 = 22.7$ in this case). While advancing the solution over every Δt, the wave elevation η is noted at the fixed r. In order to obtain a good and smooth time series, Δt is made sufficiently small and the number of time steps is increased to an average of about 300 steps per wave length. Figure 5.2 shows a time series resulting from the vertical cut.

At this stage, we recall that this is the time series resulting from a crater of $R = 10$ feet, in water depth $h = 2.7$ feet and crater lip length of 5 feet. The time series is at a distance $r_0 = 22.7$ feet from GZ. This time series could be propagated to any desired distance r from GZ using Equation (4.79) as in Chapter 4 for comparison to the recorded wave time histories obtained from real data.

2. Hydrodynamic Energy Dissipation and Effect of Lip Shape

A very important feature of this model is that it is dissipative during its early phase. While the collapse of the crater itself is not dissipative, a considerable portion of the initial potential energy is lost as a result of the turbulence in the cylindrical bore radiating outwardly from the rebound of the inward flow at GZ. This is not accounted for by the linear theory. Application of

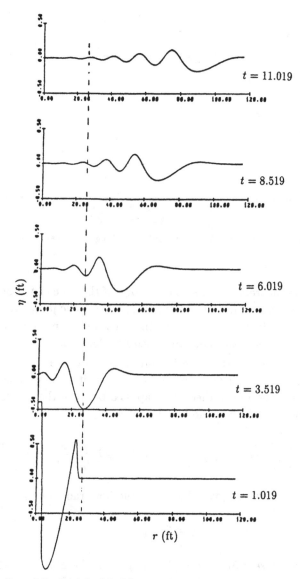

Fig. 5.1. Evolution of the initial critical bore and the leading wave using Equation (4.90). Formation of an undulated bore and then a transient wave train consisting of the leading wave and the trailing group of high frequency waves is shown. Initial condition was the previously computed critical bore and leading wave system resulting from the collapse of a crater with $R_c = 10$ feet in 2.7 feet of water (Chapter 2).

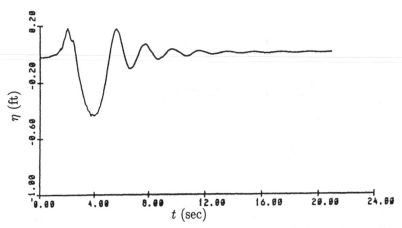

Fig. 5.2. Time series at $r = 22.7$ obtained by making a vertical cut in Figure 5.1.

the momentum theorem permits computation of the bore elevation external to the turbulent processes, just as in the case of a hydraulic jump. It is possible to compute the rate of energy dissipation in a hydraulic jump by the application of Bernoulli's theorem; the same approach is followed here.

Consider Figure 4.8; the bore formation is more like a moving hydraulic jump, and Bernoulli's theorem must be used in its full form including the unsteady terms. Bernoulli's theorem is applied between the points u and d for "up" and "down" regions of the flow and it is

$$\rho Q g \left[\frac{U_d^2}{2g} + h_d \right] - \rho Q g \left[\frac{U_u^2}{2g} + h_u \right] - \rho Q \int_d^u \frac{\partial U}{\partial t} ds = \frac{\Delta E}{\Delta t}. \qquad (5.1)$$

Q is the volume flow rate given by (R_b is the bore radius r in Figure 4.8)

$$Q = 2\pi R_b^* h_d U_d. \qquad (5.2)$$

$U_u = 0$ and $ds = W_b x \Delta t$. Thus, the integral in the above equation may be replaced by

$$\rho Q \frac{(0 - U_d)}{\Delta t} W_b \Delta t. \qquad (5.3)$$

So energy dissipation per second is given by

$$\frac{\Delta E}{\Delta t} = \rho Q g \left[\frac{U_d^2}{2g} + h_d - h_u \right] + \rho Q W_b U_d .$$ (5.4)

In this process, the effect of variation in the lip-height and lip-length on energy dissipation are noted. The main criteria for the selection of the lip size are:

- magnitude of the leading wave,
- shock formation at the leading wave.

The above two factors are conflicting in the sense that a higher lip-height would mean a higher leading wave, but a smaller lip-radius causes a spilling breaker at the leading wave. Equations to handle bore formation at the leading wave have not been incorporated. This is because of the fact that we are still uncertain about the exact shape of the lip. Figure 5.3 shows three configurations of the crater and lip dimensions. The lip-height of the crater is computed by conservation of mass as described in the previous chapter.

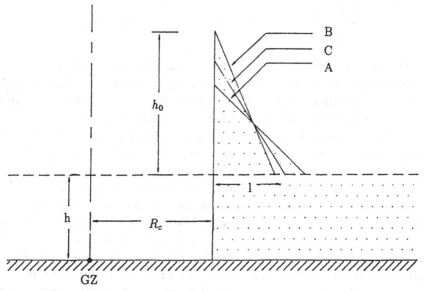

Fig. 5.3. Three trial lip shape configurations A, B, and C. The crater lip shape with the highest possible potential energy without causing the formation of breaking leading wave is chosen.

The initial potential energy in a static crater is given by the integral

$$E_0 = \rho g \int_0^{2\pi} \int_0^{\infty} \left(\frac{Z(r)}{2} \right) Z(r) r \, dr \, d\theta, \qquad (5.5)$$

where Z is the initial elevation at a distance r from GZ and the origin is on the free surface. Potential energy computations for the three lip configurations are given in Table 5.1.

Table 5.1. Potential energies corresponding to various crater lip configurations as shown in Figure 5.3.

Cases	A	B	C
Lip length	$0.75R_c$	$0.5R_c$	$\frac{8}{13}R_c$
Lip height	$\frac{16}{15}h$	$\frac{12}{7}h$	$\frac{195}{145}h$
Potential energy $2\pi\rho g h^2 R^2$	0.4175	0.5254	0.46118

The three cases listed above were subjected to rigorous numerical tests. Initial cavities varying from $R_c/h = 1.0$ to $R_c/h = 8$ were allowed to collapse and the computations were performed up to the bore limit criterion. Total energy dissipated was computed for each case (Table 5.2).

It is very interesting to note that energy dissipated is not sensitive to the lip configurations. Thus, we may choose a mass conservative lip for convenience and yet have the same amount of energy transmitted to the waves. The energy dissipated hydrodynamically remains approximately a constant 40%.

Concerning mathematical stability, models A and C were always stable but model B tended to form a shock, i.e., a breaking wave, at the leading wave. Model A resulted in very small leading and trailing waves compared to the experimental wave records, indicating that the initial energy content was not sufficient. Model C was determined by trial and error as the case which just prevents the formation of shock and allows maximum possible energy to be transmitted to the waves without causing instability. This case was selected as the optimum model for the wave predictions and calibration.

Table 5.2. Ratios of hydrodynamic energy dissipated with respect to the initial potential energies for various values of R and for three lip configurations A, B, and C.

Configurations		C	A	B
$\dfrac{R}{h}$	E_{0C}	$\dfrac{\Delta E}{E_{0C}}$	$\dfrac{\Delta E_A}{E_{0A}}$	$\dfrac{\Delta E_B}{E_{0B}}$
1.0	250.2	0.4116	0.415	0.403
2.0	1001.1	0.399	0.392	0.393
2.5	1564.3	0.399		
3.0	2252.5	0.417	0.406	0.4
3.5	3066.0	0.4		
4.0	4004.6	0.39	0.406	0.395
4.5	5068.3	0.402		
5.0	6257.2	0.404	0.414	0.408
5.5	7571.2	0.410		
6.0	9010.3	0.397	0.416	0.379
6.5	10574.6	0.390		
7.0	12264.1	0.392	0.38	0.367
7.5	14078.7	0.366		
8.0	16018.4	0.351		

3. Calibration

Model C was allowed to collapse to form a bore behind the leading wave and propagate undulatory waves using the KdV Equation (4.79), for a large number of initial conditions defined by a value of R_c (feet) over a foot of water depth. Figures 5.4 and 5.5 are examples of these computations in four different steps for $R_c = 1$ foot and $R_c = 8$ feet, respectively. Each figure shows four curves. The first curve is the initial crater. The second curve shows the collapse up to a point where the leading wave just touches GZ. The third curve shows the leading wave and the bore formation behind it at the bore limit point. This is the input to the KdV equation solver which generates the last curve showing propagating nonlinear waves. The resulting wave forms are studied.

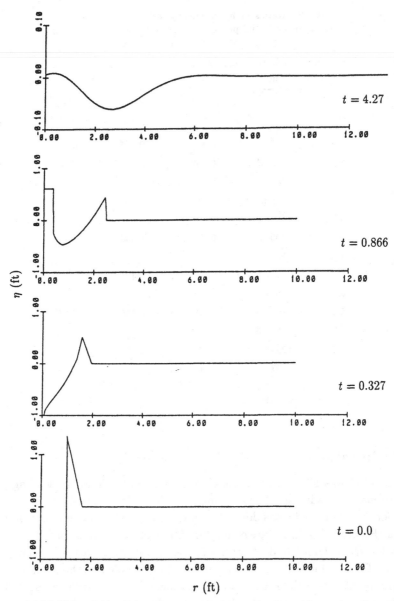

Fig. 5.4. Initial crater of $R_c = 1$ foot in water of $h = 1$ foot is allowed to collapse. Subsequent bore formation and propagation is followed by propagation of nonlinear waves using the KdV equations.

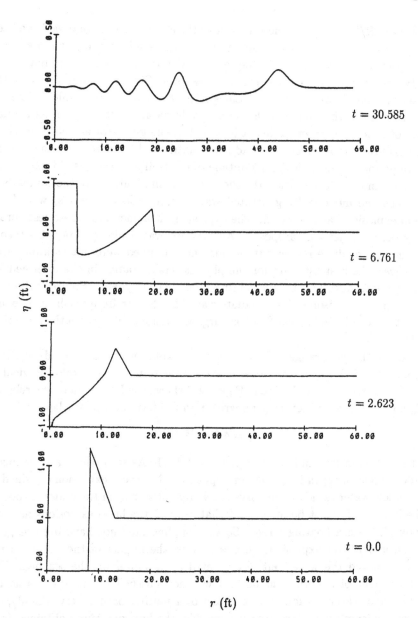

Fig. 5.5. Initial crater of $R_c = 8$ feet in water of $h = 1$ foot is allowed to collapse. Subsequent bore formation and propagation is followed by propagation of nonlinear waves using the KdV equations.

For large R/h, $[R/h > 4]$ the waves show the characteristics that we have been looking for, namely, a large leading wave, large shallow trough followed by a group of higher frequency trailing waves. It now only remains to match the data with the numerically generated wave form. Figures 5.6 and 5.7 show the resulting wave train for various values of R at a relatively equivalent time.

To achieve the match with the available data, which is given as a function of time, the crater radius R is varied at the given water depth in which the explosion took place. R is adjusted until the best match for the leading wave period, along with the magnitude of the trailing waves, is obtained. Figures 5.8 and 5.9 demonstrate the success of the method, where the comparison between the numerically generated wave records (dotted curves) with large yield explosion data is shown. The explosion phenomena for these tests probably occurs exactly as proposed. An attempt was made to apply this theory to the intermediate region but the matches obtained were not as impressive. However, the best estimate for the physical crater radius in the intermediate water depth also may be given using modeling judgment. Figures 5.10–5.12 show the comparison of intermediate water depth data from small scale explosion tests conducted at WES, Vicksburg, Mississippi, USA, with the numerical solution.

Using the wide range of data from small explosions (a few pounds of TNT) to large yield (KT of TNT), a calibration curve to relate the explosion yield to the crater radius was obtained. Figure 1.7 shows the least squares fit relating R_c and W. The relation may be written in the form of a power law as

$$R_c = 4.3929 W^{0.2476} \qquad (5.6)$$

where R_c is in feet and W is in pounds of TNT. As seen, the crater radius is independent of depth based on our hypothesis that every explosion classified as a shallow water explosion removes the water entirely from the crater, exposing the bottom. It must be mentioned that some of the large explosions actually took place on a floating barge. So the coupling may not have been as good as an underwater explosion. In many cases, the impact of the explosion not only removed the water but also created a big crater on the sea bed. The requirement of a horizontal bottom was not satisfied in many cases as the waves had traveled a few thousand feet over natural bathymetry. Finally, the measurements themselves are questionable, the best one being obtained from the movement of ships moored at a distance from GZ. Taking into account the errors introduced due to the above-mentioned shortcomings, one may retain

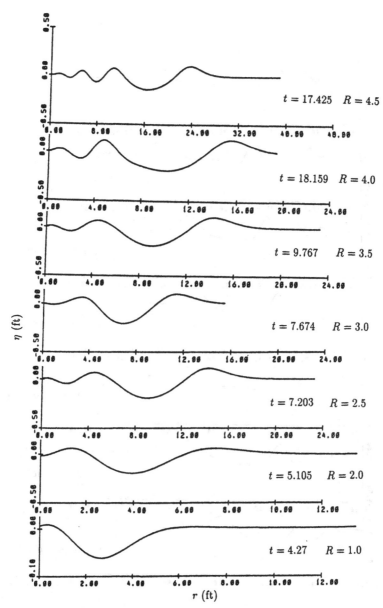

Fig. 5.6. Free surface elevation at relatively equivalent time as a function of the initial radius of the crater R_c from $R_c = 1$ to $R_c = 4.5$ feet using the KdV equations (water depth $h = 1$ foot).

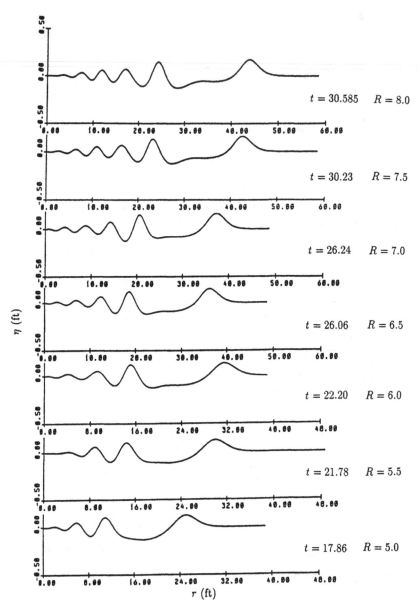

Fig. 5.7. Free surface elevation at relatively equivalent time as a function of the initial radius of the crater R_c from $R_c = 5$ to $R_c = 8$ feet using the KdV equations (water depth $h = 1$ foot).

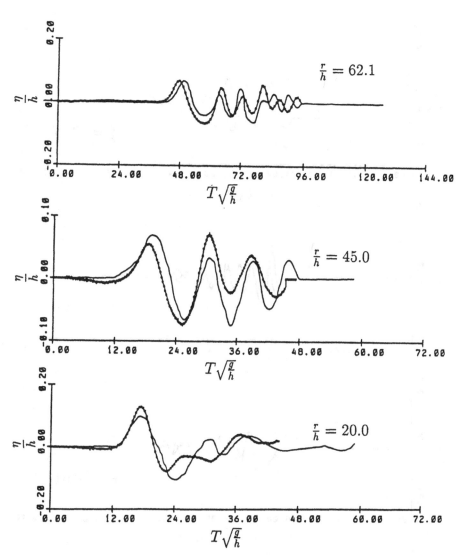

Fig. 5.8. Comparison of wave records resulting from large yield shallow water explosions with the numerically generated wave records using the nonlinear wave theory with from top to bottom $R_c/h = 5.4$, 2.7, 3.3. The near-field time series at bore collapse is taken at $r/h = 19.09$, 14.89, 8.26 respectively, and the wave records are at $r/h = 62.1$, 20, 45. The dotted curves are the theoretical predictions.

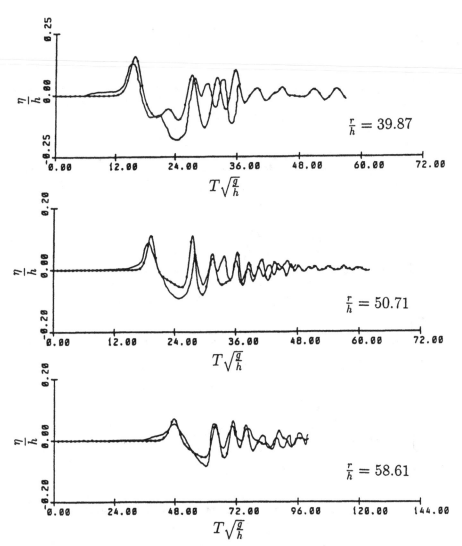

Fig. 5.9. Comparison of wave records resulting from large yield shallow water explosions with the numerically generated wave record using nonlinear wave theory from top to bottom $R_c/h = 7$, 5.7, 4.5 respectively. The near-field time series at bore collapse is taken at $r/h = 19.37$, 14.36, 11.58 respectively, and the wave records are at $r/h = 39.87$, 50.71, 58.61. The dotted curves are the theoretical predictions.

(a)

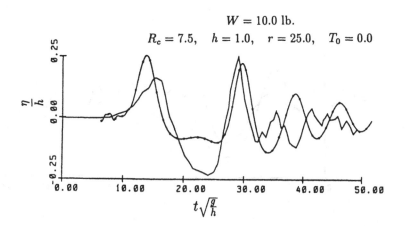

(b)

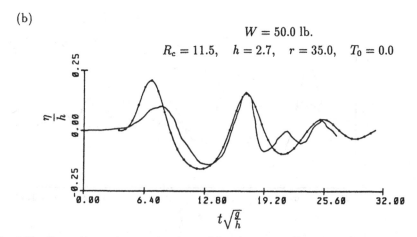

Fig. 5.10. Comparison of wave records resulting from a small yield shallow-intermediate water explosion with the numerically generated wave record using the nonlinear theory. The dotted curve is the theoretical prediction. (a) $W = 10$ pounds, $h = 1.0$ feet, $r = 25.0$ feet. (b) $W = 10$ pounds, $h = 2.7$ feet, $r = 27.7$ feet.

(a)

$W = 20.0$ lb.

$R_c = 8.5, \quad h = 2.7, \quad r = 25.0, \quad T_0 = 0.0$

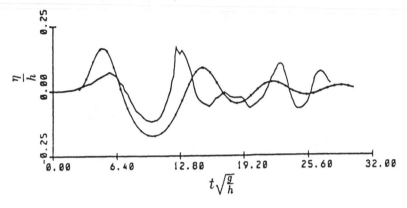

(b)

$W = 30.0$ lb.

$R_c = 10.5, \quad h = 2.7, \quad r = 30.0, \quad T_0 = 0.0$

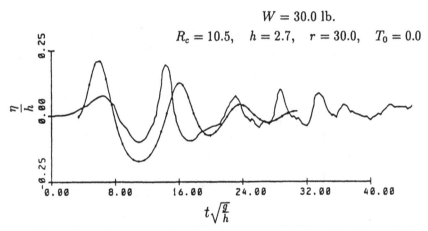

Fig. 5.11. Comparison of wave records resulting from a small yield shallow-intermediate water explosion with the numerically generated wave record using the nonlinear theory. The dotted curve is the theoretical prediction. (a) $W = 20$ pounds, $h = 2.7$ feet, $r = 25.0$ feet. (b) $W = 30$ pounds, $h = 2.7$ feet, $r = 27.7$ feet.

(a)

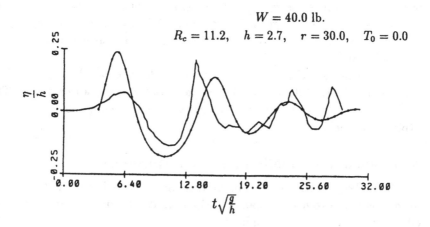

(b)

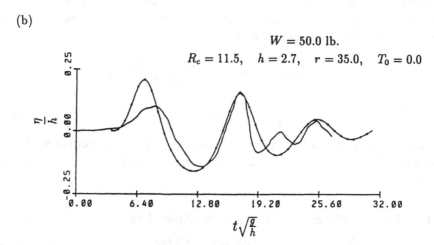

Fig. 5.12. Comparison of wave records resulting from a small yield shallow-intermediate water explosion with the numerically generated wave record using the nonlinear theory. The dotted curve is the theoretical prediction. (a) $W = 40$ pounds, $h = 2.7$ feet, $r = 30.0$ feet. (b) $W = 50$ pounds, $h = 2.7$ feet, $r = 35.0$ feet.

$R_c = 4.4W^{0.25}$ for the predictions after rounding off the decimals. Only one measurement of the real crater resulting from a 1-pound explosion in 1 foot of water has been made using high-speed photography. The measured physical crater fits the same law given by Equation (5.6) which further confirms the theory. Figure 1.7 shows also the measured crater radius corresponding to a 1-pound charge.

Thus, given a yield W, the crater radius R_c can be obtained by Equation (5.6). Model C defines the lip parameters using the water depth. Then collapse of this crater results in waves that closely resemble waves from a real explosion of yield W.

4. Efficiency of Shallow Water Explosions

Since a large portion of the potential energy imparted to the water in crater formation is lost to hydrodynamic dissipation; it is interesting to calculate the efficiency of shallow water explosions.

The yield of an explosion or the energy released in the explosion process is described in equivalent pounds of TNT. Since 1 pound of TNT $\approx 1.9 \times 10^6$ J, the energy yield can be expressed in Joules. For model C in water depth h, the potential energy in the crater W_1 is:

$$W_1 = 2\pi\rho g(0.46118)\left(\frac{R_c}{3.28}\right)^2\left(\frac{h}{3.28}\right)^2, \qquad (5.7)$$

which is

$$W_1 = 251.48 h^2 R_c^2 \text{ Joules}, \qquad (5.8)$$

where h and R are in feet. We have demonstrated that hydrodynamic loss is $\approx 40\%$ (Table 5.2). If W_2 is the energy contained in the propagating waves, then

$$W_2 = 0.6W_1. \qquad (5.9)$$

Therefore, the efficiency ε is now directly obtained as

$$\varepsilon = \frac{W_2}{W_1} \times 100 = \frac{\text{Energy in Waves}}{\text{Energy Released in Explosion}} \times 100. \qquad (5.10)$$

Using Equations (5.6), (5.8), and (5.9), the efficiency may be expressed as

$$\varepsilon = 0.1529\left[\frac{h}{W^{0.2524}}\right]^2, \qquad (5.11)$$

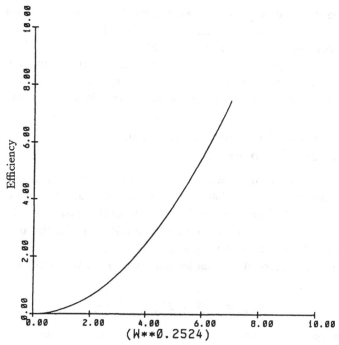

Fig. 5.13.   Efficiency of shallow water explosions as a function of yield W and water depth h.

where h is in feet and W is in pounds of TNT. Figure 5.13 shows a plot of efficiency versus $[h/W^{0.2524}]$. Table 5.3 shows examples of efficiencies of typical shallow water explosions.

Table 5.3. Examples of typical shallow water explosion efficiencies.

W in KT	Depth h (ft)	ε (%)
1	12	0.0137
1	63	0.3812
1	126	1.524
1000	126	0.0465
1000	630	1.666
1000	1259	4.658

It is seen that shallow water explosions are extremely inefficient with respect to wave generation ability. Typically, at most only $\approx 5\%$ of the explosion energy is obtained in the form of waves. Most of the energy is distributed in other explosion effects. A part is lost to the shock wave in water and in the atmosphere via the free surface. Of the remaining energy, 40% is lost due to hydrodynamic dissipation and the remaining energy is used in wave-making.

5. Comparison Between Linear and Nonlinear Theories

The importance of linear wave theory as a practical tool in deep and intermediate depth waters has been established. The same model extended to shallow water predictions succeeded in producing large leading waves. Using the leading wave amplitude as the governing criterion, the linear model was calibrated for shallow water (Wang *et al.*, 1991).

The linear wave record is given by Equation (3.25) where the scaling coefficient

$$\eta_{\max} = \frac{\eta_{\max}^*}{h^*} = 2.7(R)^{-0.73}\frac{W^{\frac{1}{4}}}{h^*} . \tag{5.12}$$

In other words, the vertical scaling parameter solely depends on the yield of the explosion. The nondimensional crater size R is given by

$$R = \frac{R^*}{h^*} = 30.0 - 24.9\left(\frac{h^*}{W^{\frac{1}{3}}}\right)^{0.07} . \tag{5.13}$$

The above empirical relations were obtained by calibration with a large data set of the maximum wave amplitudes recorded. It was established that in shallow water, the model parameters are independent of the depth of burst effect.

The result of the nonlinear theory is $R_c \propto W^{\frac{1}{4}}$ (5.6). The depth of burst effect was not considered for the same reason that the wave response was practically constant for different depths of burst in shallow water. The effect of water depth is included in the calibration of the mass conservative crater.

For a comparison of the wave prediction ability of the two theories, the Mono Lake explosion was chosen. Here, $W = 9250$ pounds, and $h = 10$ feet. The linear theory parameters using Equations (5.12) and (5.13) are $\eta_{\max} = 0.686$ and $R_c = 6.359$. The nonlinear wave theory prediction of the crater radius using Equation (5.6) is $R_c = R \times h = 53$ feet.

The two theories were applied and the resulting wave forms at $r = 621$ feet are plotted on the same graph (Figure 5.14) along with the data recorded. The curve with circles shows the nonlinear prediction while the triangles represent the linear theory. The smooth curve is the real wave record.

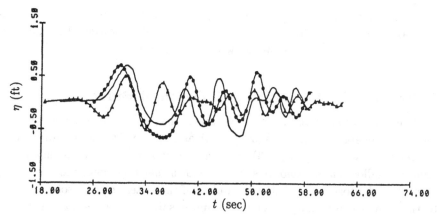

Fig. 5.14. Comparison of the linear theory and the nonlinear theory applied for the prediction of shallow water waves. The Mono Lake experiment is used for the qualitative and predictive comparison. The data from the explosion of $W = 9250$ pounds, recorded at $r = 621$ feet at a depth of 10 feet. The curve with circles represents the nonlinear prediction, the curve with triangles is the linear prediction while the data is given by the smooth curve.

Clearly, both theories accurately predict the leading wave. However, the expected shortcoming of the linear theory is its inability to give the correct nonlinear wave form. Linear theory is unable to predict the characteristic long shallow trough. Thus, the leading wave period as per linear theory is too small, nearly 50% off. This affects the representation of the following trailing waves. Similar results may be expected for other cases in this shallow nonlinear region. Therefore, we conclude that the proposed nonlinear theory correctly describes the phenomena and predicts the waves both in magnitude and shape with better accuracy than the linear theory.

In summary, when the relative water depth $d/W^{1/3}$ is less than unity, a method has been developed which allows the determination of the physical size of the water crater which reaches the seafloor. The crater was chosen schematically as being with vertical limit and surrounded by a triangular lip. Its exact shape has not been determined but it does not seem to influence the resulting wave.

The method seems to provide a good estimate of the energy dissipated by turbulence even though the energy dissipated by the plume is not exactly known. Finally, the method allows the determination of the nonlinear features of the wave train and in particular the time elapsed between the first two wave crests.

Accordingly, given the yield W and the water depth, one is now able to estimate the wave motion at any distance from GZ following the explosion including to some extent the region within the crater.

6. Effect of Water Depth and Depth of Burst

The calibration of the linear wave theory has indicated a dependency of the wave generation processes in general and the linear crater radius R in particular to water depth (5.13) as seen in Chapter 3. This dependency is mild for shallow water explosions. It was shown that these dependencies could be neglected altogether and that the radius of the water crater is solely related to the yield. The validity of such a simple result may be questioned, but it seems to hold true within the accuracy of the experimental results. However, it is reasonable to assume that more controlled experiments would indicate that both the water depth and the depth of burst affect the size of the water crater. This trend is particularly significant in the case of the depth of burst z.

Even though the experimental results obtained at WES are erratic to indicate a definite trend, it seems that the wave amplitude of the maximum wave suddenly decreases by half when the depth of burst tends to zero, i.e., when the charge is at the free surface. The nuclear results seem to yield the same trend, even though the depth of bursts are ill-defined when the charges were aboard floating barges.

There is an indication that the radius of the water crater decreases slightly when the crater depth decreases, as indicated by a recent experiment done in very shallow water at WES, with very small yield (0.25, 0.50 pounds of TNT) behind a glass window (Bottin, 1990). A high-speed camera provided the time history of the water crater formation and its collapse (Figure 5.15). The results support the theoretical developments presented in Chapter 3 for the most part (the presence of a steel plate near GZ does not allow verification of the theory of bore formation decay and transformation into a KdV wave).

The analysis of the wave record via the theory presented in Chapter 4 would allow us a direct comparison of the theoretical crater R_c versus the

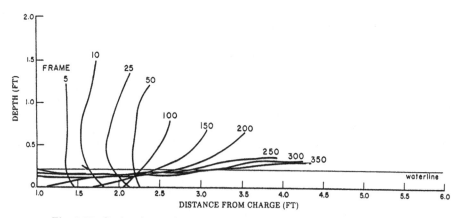

Fig. 5.15. Cavity shapes due to explosion in shallow water (Bottin, 1990).

experimentally observed one. This work has not yet been done. The tests indicate that the maximum crater size is slightly smaller than predicted by our previous analysis (by about 5–15%) and that the discrepancy increases as the water depth (or depth of burst) decreases.

To resolve this problem, more tests are needed. Nevertheless, until more carefully planned tests are done, the overall conclusion of this chapter, which assumes that only the yield matters, holds true and the accuracy of our method of prediction is sufficient for most purposes.

Chapter 6

DISSIPATIVE PROCESSES OF WAVE PROPAGATION

1. Importance of Dissipative Processes by Wave-SeaFloor Interaction

Dissipative processes have an important effect on all the phases of generation and propagation of EGWWs. During the generation phase, the amount of energy dissipated by turbulence in the cylindrical bore resulting from the collapse of the water crater in shallow water can be determined theoretically. It was found that about 40% of the potential energy of the original crater was lost in the process. Once the wave radiates, the dissipation processes continue as a result of wave-seafloor interactions and wave breaking.

In the case of large yields exploding in deep water ($W > 1$ MT), the dissipative process by wave-seafloor interactions become important when the waves reach the depths encountered on typical continental slopes. There are two reasons why this dissipative process is important, as compared to wind waves and tsunami waves.

First, EGWWs of interest have a much longer wave length than wind waves, therefore the wave motion interacts with the seafloor at deeper depths than wind waves. For example, a 44-second wave, which is the period of the maximum wave resulting from a 1-MT explosion in deep water, starts to interact

217

with the seafloor at depths as deep as 5,000 feet and it becomes a shallow water wave when the water depth is shallower than 100 feet ($\frac{d}{L} < 0.04$). By comparison a 12-second wind wave "feels" the sea floor at a depth of 300 feet and becomes a shallow water wave at a depth of only 7.4 feet.

The second reason wave-seafloor interactions are important is amplitude. A tsunami wave interacts with the seafloor in the deepest ocean but its amplitude is small, say usually one foot in deep water, which grows to less than 2 feet at 300 feet at the edge of the shelf. (Tsunami waves are partially reflected on the continental slope prior to reaching the continental shelf (Le Méhauté, 1971).) The turbulent dissipation process is nonlinear. The shear on the seafloor is grossly proportional to the square of bottom velocity, u_b^2. The energy dissipated is proportional to u_b^3. According to the shallow water linear wave theory, the amplitude of bottom velocity u_b is:

$$u_b = \frac{H}{2}\sqrt{\frac{g}{d}}, \tag{6.1}$$

which is independent of the wave period and directly proportional to the wave height H.

Therefore, the dissipated energy is proportional to H^3. It is recalled that the energy flux is proportional to H^2. Therefore, the wave decaying coefficient is proportional to the wave height H: The larger the wave height, the larger the wave damping. It is seen that an EGWW with a wave height in a range of say 100 feet, is damped 50 times faster than a tsunami wave of a longer period.

The damping of EGWWs is much larger than both the shorter wind waves and the longer tsunami waves, thus, the damping of EGWWs over a long continental shelf, far from being negligible, is a necessary fact to take into account in any realistic prediction on the effects of EGWW. When EGWWs arrive on a sloped beach and break, the dissipation process is further enhanced as was explained in Chapter 1.

The damping of EGWWs by wave-seafloor interactions is a function of the soil characteristics. For example, on the East Coast, the seafloor is made mostly of sand and the prevailing dissipative process is in the turbulent boundary layer over a moveable bed made of sand. Near the gulf, the soil is made of ooze, fine silt, and mud so that the soil responds to wave pressure-induced fluctuations. The soil response induces a dissipative Coulomb friction between soil particles, which is the prevailing wave dissipation processes. The loss of energy by pressure-induced percolation through the soil may also intervene.

2. Conservation of Energy Flux

Based on what has been learned from wind waves, there is now enough evidence indicating that dissipative processes may significantly modify the height of EGWWs on long continental shelves such as that encountered on the East Coast of the United States.

The basic approach taken herein is initially to treat the waves as linear and monochromatic as a first approximation. Extensions of the model to EGWWs can then be obtained by following a wave element of constant frequency, as will be described in the next chapter. The traditional approach is based on the conservation of energy flux along a wave ray which is mathematically expressed by:

$$\frac{d(bF_{av})}{ds} = -b\phi_D \qquad (6.2)$$

where

- b is a distance between wave orthogonals,
- s is a wave ray,
- ϕ_D is the dissipation function which expresses the average amount of energy dissipated per unit area,
- F_{av} is the average wave flux which can be expressed for linear or nonlinear waves.

In accordance with the traditional approach it will be assumed that the dissipation function ϕ_D is the sum of the dissipation functions due to various mechanisms without interaction.

It is known that many dissipation mechanisms by seafloor interactions do exist. It can be shown that, prior to breaking, the dissipation processes in the free field are generally negligible (Mei, 1983; Chapter 8). Therefore, the most important dissipation processes are identified as resulting from

(1) the shear between wave-seafloor taking place in the (turbulent) boundary layer,
(2) the energy losses by percolation in the underlying soil under wave pressure fluctuations,
(3) the soil response under these pressure fluctuations involving Coulomb friction between soil particles.

The dissipation functions resulting from these three processes are labeled ϕ_f, ϕ_p, and ϕ_s respectively, such that

$$\frac{d(bF_{av})}{ds} = -(\phi_f + \phi_p + \phi_s)b \,. \tag{6.3}$$

It is assumed that there is no interaction between the first process and the remaining two and that these groupings can be determined independently of each other. The first function ϕ_f is now established for a horizontal seafloor and periodic waves. It is assumed that the formulation is valid locally for transient waves (EGWWs) over a nonuniform bathymetry with gentle slopes.

3. Turbulent Boundary Layer Effects

As EGWWs propagate in shallow zones, boundary layers are established as the near bottom velocity field associated with these waves must be transformed within a very small length scale to meet the conditions of zero tangential velocity at the boundary. The existence of velocity gradients normal to the direction of the flow is known to be a consequence of shear stresses acting in the fluid. For example,

$$\tau_{xz} = (\mu + \varepsilon_D)\frac{\partial u}{\partial z} \,, \tag{6.4}$$

where μ is the viscosity of the fluid, ε_D is the eddy viscosity of the fluid that exists only for turbulent flow, and u is the particle velocity in the x direction in the boundary layer.

While conceptually valuable, the above equation is usually not satisfactory for determining the shear stresses in EGWWs because the velocity profile in the boundary layer is very difficult to define. Under turbulent flow conditions, which predominantly occur in the boundary layers, sand motion is initiated and ripples of varying geometry may be formed. The sand grain sizes and the complex ripple geometry confound attempts to adequately prescribe velocity profiles in the boundary layer.

An alternate scheme is typically employed to determine the shear stress at the bottom boundary. Equation (6.5) relates the flow immediately above the boundary layer to the shear stress at the bed through a factor f commonly called the Darcy-Weisbach friction factor:

$$\tau_b = \frac{1}{2}\rho_f f u_{bm}^2 \cos(\sigma t + \xi)|\cos(\sigma t + \xi)| \tag{6.5}$$

where

- τ_b is the shear stress at the bed,
- ρ_f is the fluid density,
- ξ is the angle by which the shear velocity u^* and the bottom particle velocity u_b are out of phase, and
- u_{bm} is the maximum bottom velocity just outside of the boundary layer.

This equation is usually employed for rough turbulent and transitional boundary layer regimes. (For laminar flows, a simpler model can be used.)

Putnam (1949) has suggested that the instantaneous rate of energy dissipation ϕ_f in the turbulent boundary layer per unit area can be given by the relation:

$$\phi_f = \tau \cdot \bar{u}_b = \frac{1}{2}\rho_f f u_{bm}^3 \cos(\sigma t + \xi)|\cos(\sigma t + \xi)|\cos(\sigma t). \tag{6.6}$$

If u_{bm} is determined from the linear wave theory, this relation can be expressed as:

$$\phi_f = \frac{1}{2}\left|\frac{H\sigma}{2\sinh(kd)}\right|^3 \rho f \cos(\sigma t - kx + \xi)|\cos(\sigma t - kx + \xi)|\cos(\sigma t - kx), \tag{6.7}$$

where H denotes the wave height, k the wave number, σ the frequency, t the time, x the axis along which waves propagate, and d the undisturbed water depth.

When averaged over an appropriate time interval, the average rate of energy dissipation in the boundary layer per unit area is given by the dissipation function of Equation (6.8):

$$\bar{\phi}_f = -\frac{2}{3}\frac{\rho H^3 \pi^2 f}{T^3 \sinh^3(kd)}F(\xi), \tag{6.8}$$

where the phase correction factor $F(\xi)$ provides a correction to the rate of energy dissipation, typically ranging in value from unity to 0.75. In practice, the correction factor is usually omitted as will be the case of EGWWs.

Consider a two-dimensional wave ($b = $ constant). The energy conservation equation for this simplified case can be written as:

$$\frac{dF_{av}}{ds} = -\bar{\phi}_f. \tag{6.9}$$

Integration of this equation between two distances s_1 and s_2 yields

$$F_{av_2} - F_{av_1} = -\int_{s_1}^{s_2}\phi_f ds. \tag{6.10}$$

Substitution for the energy flux given by the linear wave theory and dissipation function yields:

$$H_1^2 - H_2^2 = \frac{16\pi^2}{3gnLT^2 \sinh^3(kd)} \int_{s_1}^{s_2} fH^3 ds\,, \qquad (6.11)$$

where $n = 1/2(1 + (2kd)/\sinh(2kd))$. This equation is nonlinear and difficult to solve accurately. A simplifying scheme is to make ds small such that the wave height H and the friction factor f remain essentially constant over the interval. In this manner Equation (6.11) can be approximated by

$$H_2^2 - H_1^2 \approx \frac{16\pi^2 f_1 H_1^3 \Delta s}{3gnLT^2 \sinh^3(kd)}\,. \qquad (6.12)$$

It is noted that the wave height H_1 and friction factor f_1 at the beginning of an interval Δs are used to compute the rate of damping throughout the interval. Dividing Equation (6.12) by H_1^2 leads to

$$H_{t_2} \approx H_{t_1}\left(1 - \frac{16\pi^2 f_1 d_{t1} \Delta s}{3gnLT^2 \sinh^3(kd)}\right)^{1/2}\,. \qquad (6.13)$$

We now make the following substitution:

$$2D_B = \frac{-16\pi^2 f_1 H_1}{3gnLT^2 \sinh^3(kd)}\,. \qquad (6.14)$$

We observe in Equation (6.14) that when $2D_B\Delta s \ll 1$, the equation can be closely approximated by the relation

$$H_2 = H_1 \exp[+D_B\Delta s]\,. \qquad (6.15)$$

Equation (6.15) approximates Equation (6.13) to within 1% error when the product $D_B\Delta s$ is less than 0.10.

To minimize the error associated with this approximate method of calculating reduced wave heights it is suggested that the interval Δs be on the order of one kilometer. When the wavelength is longer than one kilometer, it is recommended that the wave length be taken as the incremental distance Δs along a wave ray.

It should be noted that this method slightly overestimates the rate of wave damping over flat and mildly sloped bottoms (i.e., $dd/ds \leq 0.005$) since the rate of damping is assumed to be constant over an incremental distance when in actuality it decays in proportion to the wave height over the interval. Qualitatively, this method becomes more accurate for bottom slopes that are steeper since the damping and shoaling effects tend to offset, keeping the wave height, and thus the rate of damping, more constant over the interval as assumed. When Δs is taken too large, however, the approximation of Equation (6.13) by Equation (6.15) tends to underestimate the rate of damping.

Nevertheless, this procedure poses a very tractable method for calculating wave height transformation due to energy dissipation, except that f is a semi-empirical coefficient which depends on the bottom flow regime which remains to be determined. The following section is, therefore, devoted to defining the bottom profile under various wave conditions, and thereby defining the friction factor f.

4. Wave Friction Factor Over a Moveable Bed — Overview

The wave friction factor f over a fixed and moveable bed has been determined, starting with the work of Jonsson (1966). A good synthesis of this work is presented by Graber and Madsen (1988). Theoretical expressions for f were obtained by Kajiura (1964 and 1968) and Grant (1977) which introduced the concept of eddy viscosity equations. Trowbridge and Madsen (1984) improved on the formulation of f by including more realistic eddy viscosity models. In general, the wave friction factor is found to be a slowly varying function of the parameters characterizing the flow and the bottom roughness. A detailed treatment of this subject can be found in Grant and Madsen (1982).

Grant and Madsen (1986) also argued that the fluid-sediment interaction must be accounted for as precisely as possible for accurate predictions of wave energy loss due to bottom friction (and this would certainly be the case for EGWWs). From analyses of laboratory data, they developed a formula for the equivalent bottom roughness of a moveable bed under the action of waves. Their roughness, which depends on sediment and near-bottom wave characteristics, may in turn be used in conjunction with their model, to predict friction factors for a moveable bed in the presence of waves. Figure 6.1 presents the wave friction factors, f, as a function of the near-bottom wave orbital excursion amplitude, A_b, for various diameter quartz sands. Although the predicted values of f depend on wave period, this dependency is sufficiently weak to regard the results presented in Figure 6.1 for a period of 10 seconds to be representative of the range of wind wave periods expected in the marine environment. In the case of EGWWs with a longer period, Figure 6.1 is not quantitatively valid and the wave friction factor needs to be recalculated.

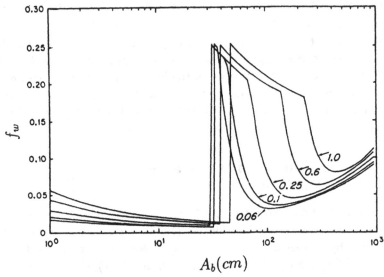

Fig. 6.1. Moveable bed friction factor f_w vs. near-bottom orbital excursion amplitude, A_b, obtained from Graber and Madsen (1988) for a 10-second wave as a function of diameter (mm) of a quartz sand.

Physically, the friction factor variation shown in Figure 6.1 represents, for low values of A_b, the flow resistance of a flat, immobile bed, i.e., with a roughness equal to the sediment diameter. As A_b increases, the wave-sediment interaction reaches a threshold value (Madsen and Grant, 1976) at which sediment starts to move. Once the threshold value is exceeded, the sediment-fluid interface deforms and wave-generated ripples appear. Initially, this is the "equilibrium regime" characterized by the formation of alternate vortex on both sides of the sand ripple crest. The bottom roughness now scales vortex with the ripple geometry, which results in a dramatic increase in friction factors. Further increase of the flow intensity, i.e., A_b, "smoothens" the ripples, first gradually then rapidly, and results in corresponding rates of decrease of the friction factor with increasing A_b until the ripples are practically washed out and the bed is returned to its initial flat state. This is the "post vortex regime". The friction coefficient reaches a minimum value which is larger than the value associated with the initial flat bed. As A_b increases further, in contrast with the immobile flat bed state, sediment now moves back and forth above the bed. This is the "sheet-flow regime" associated with the sediment in motion. There is an apparent increase in roughness reflected by the friction factor for large A_b values. Beyond the minimum of the friction factor for a mobile flat bed its variation is controlled by the increase in intensity of sediment movement above the bed. In many cases, EGWWs are in the "sheet flow regime", moving a considerable amount of sediment across the seafloor.

5. Procedure for Determining the Wave Friction Factor for EGWWs

Determining the wave friction factor over a moveable bed is very complex. It has been described thoroughly for EGWWs by Colby (1984), and will not be repeated here. The justification and basis for the selected formulation is given in the aforementioned report, which is a direct adaptation of the work of Grant and Madsen (1986) on EGWWs.

The determination of the wave friction factor over a moveable bed requires a succession of calculations and iterative procedures:

(1) Initially, the horizontal fluid displacement immediately above the boundary layer is determined using the linear wave theory, i.e.,

$$A_b = \frac{\eta_E}{\sinh kd}, \qquad (6.16)$$

where η_E is the amplitude of the wave envelope, k is the local transiential wave number, and d is the undisturbed water depth. The maximum bottom velocity at the limit of the boundary layer is:

$$u_b = A_b \sigma, \qquad (6.17)$$

where σ is the transiential frequency ($\sigma = \frac{2\pi}{T}$).

(2) Next, verify that the flow at the boundary is turbulent, which occurs when the Reynolds number is:

$$\frac{u_b \delta}{\nu} > 160, \qquad (6.18)$$

where δ is representative of the boundary layer thickness,

$$\delta = \left(\frac{2\nu}{\sigma}\right)^{1/2},\tag{6.19}$$

and ν is the kinematic viscosity.

(3) Compute whether or not grain motion is initiated. This occurs if $u_b > u_c$. u_c is a conceptual bottom velocity which depends on the flow regime. For EGWWs, the regime is generally completely mixed, which occurs when the eddy viscosity $\varepsilon_D \geq 1$; therefore,

$$\frac{A_b}{D_s} \leq 355\left(\frac{\nu}{\sigma}D_s^2\right)^{0.645},\tag{6.20}$$

where D_s is the sediment diameter. Then

$$u_c = (8\gamma'gD_s)^{1/2},$$
$$\gamma' = SG - 1,\tag{6.21}$$

where SG is the grain specific gravity.

(4) Evaluate the ratio of the hydrodynamic drag force on the grain to its weight θ_D:

$$\theta_D = \frac{\rho u_b^2}{\gamma_s D},\tag{6.22}$$

where

- ρ is the fluid density,
- γ_s is the relative specific gravity of the sediment $(\rho_s - \rho)g$, and
- ρ_s is the sediment density.

(5) If $\theta_D \leq 40$, the equilibrium vortex regime prevails, then

$$f = 0.23k_s^{0.9},\tag{6.23}$$

and the relative hydraulic roughness is

$$k_s = \frac{2D_{90} + 0.01\eta_r}{A_b},\tag{6.24}$$

where D_{90} is the diameter of the largest sediment (90%) and η_r is the sand ripple height which is given by

$$\eta_r \cong 0.2A_b,\tag{6.25}$$

(the ripple length $\lambda_r = 4/3a_b$).

(6) If $40 < \theta_D < 240$, the post vortex regime prevails. Then the sand ripple geometry is defined by

$$\frac{\eta}{\lambda_r} = \exp\{-1.113 * 10^{-8}\theta^4 + 1.846 * 10^{-6}\theta^3$$

$$+ 3.423 * 10^{-4}\theta^2 - 8.144 * 10^2\theta + 0.933\}, \tag{6.26}$$

where λ_r is the length of the sand ripple defined by

$$\frac{\lambda_r}{D_s} = 10^2\{3.742 \times 10^3 p^4 - 0.1435 p^3 + 1.970 p^2 - 11.436 p + 29.6911\}, \tag{6.27}$$

where $p = (\frac{a_b}{D_s})10^{-3}$ and D_s are the median diameter.

(7) If $\theta > 240$, it is the sheet flow regime and all ripples are effaced.

(8) In the post vortex and sheet flow regime, the thickness of the saltation layer h_t is determined:

$$h_t = 42.3(SG + C_m)D\psi_c\left[\left(\frac{\psi'}{\psi_c}\right)^{1/2} - 0.7\right]^2. \tag{6.28}$$

Also

$$\psi' = \frac{f'u_{bm}^2}{2(SG - 1)gD}, \tag{6.29}$$

and $f' = f$ is given by the following equation:

$$\frac{1}{4.05 f^{1/2}} + \log\frac{1}{4f^{1/2}} = -0.254 + \log\frac{a_b}{k_N}, \tag{6.30}$$

where k_N is equal to D_s.

Finally, ψ_c is given as a function of

$$S_* = \frac{D}{4\nu}\sqrt{(SG - 1)gD}, \tag{6.31}$$

by Figure 6.2.

(9) Once h_t is determined, the relative hydraulic roughness equivalent k_s which takes into account both the form drag due to sand ripples (when they exist) and the near-bed sediment transport is

$$k_s = \frac{160}{a_b}(SG + C_m)D\psi_c\left[\left(\frac{\psi'}{\psi_c}\right)^{1/2} - 0.7\right]^2 + 28\frac{\eta_r}{a_b}\frac{\eta_r}{\lambda_r}, \tag{6.32}$$

where C_m is the added mass coefficient of sediment particle.

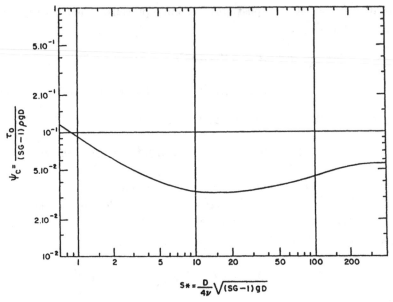

Fig. 6.2. Modified shields diagram for the initiation of sediment movement (Madsen and Grant, 1976).

(10) Then the friction coefficient f can now be determined by successive approximations by the equation:

$$f = \frac{\kappa^2}{2}(ker^2 2\zeta_0^{1/2} + ker^2 2\zeta_0^{1/2})^{-1} , \qquad (6.33)$$

where κ is the von Karman constant (0.4) and ker and kei are the Kelvin functions of zeroth order; $ker(x)$ and $kei(x)$ are the respective real and imaginery parts of $K_0(xe^{-\pi/4})$. The Kelvin functions of zero order are computed using the approximations given by Abramovitz and Stegun (1964) (see Chapter 9, Sections 11.3 and 11.4) as functions of x. Here:

$$x = 2\zeta_0^{1/2} . \qquad (6.34)$$

Finally,

$$\zeta_0 = \frac{\sqrt{2k_s}}{30\kappa f^{1/2}} . \qquad (6.35)$$

Note, f appears in ζ_0, which is why f has to be determined by successive approximations by an iterative scheme (start with $f \cong 0.01$).

(11) Once f is determined, the damping coefficient D_B (6.14) can be determined for an interval Δs.

Tabulated values of damping coefficients D_B, as defined by Equation (6.14) due to energy dissipation in the turbulent boundary layer are presented for varying values of water depth, wave height, period, and sand grain size in Table 1.14. They are presented in more detail in Tables 6.1, 6.2, and 6.3. It should be noted that the wave heights and periods utilized in the tables are representative of those anticipated in explosion-generated waves. An application of these coefficients has been presented in Chapter 1 in computing the actual damping of EGWWs over a long shelf.

As a rough check on the validity of the friction coefficients computed by the formulation presented in this section, we can compare the friction factors obtained in the wind wave range to those actually obtained by a number of investigators in the field. It must be realized that, in general, the field investigators were dealing with random sea states rather than monochromatic waves as dealt with herein. Thus, to obtain friction coefficient for the random sea states, a methodology such as that presented by Hasselman and Collins (1968) may have been employed. The computed friction factors are shown in Table 6.4 and the experimental field values in Table 6.5.

It is seen from a comparison of the tables that, in general, the computed and experimental friction factors are on the same order of magnitude, though the computed values are usually slightly smaller. Neglecting all of the other uncertainties in the experimental data, this may be explained by the fact that dissipation by percolation and sea-seabed response actually occurred and was implicitly taken into account in the friction factors. With this in mind, the good comparison in the wind wave range suggests that our formulation presented here is indeed reasonable, and that its extension to the range of EGWWs should also give reasonable estimates of the damping that would actually occur. Nevertheless, one must keep in mind that only four experimental points are available in the open literature (Carstens *et al.*, 1969) in the sheet flow regime where most applications to EGWWs lie. Furthermore, there are no existing natural phenomena where the shear on the seafloor could be as high; for example, one could have a bottom velocity of 10 m/sec. at the limit of a 1- or 2-meter thick boundary layer. Therefore, a doubt remains on the value of our quantitative extrapolations to EGWWs. But one knows for sure that this effect is important even though ill-defined.

Table 6.1. Dissipation by the turbulent boundary layer fine sand (by USC Standard). $D_s = 0.1$ mm, $SG = 2.65$ damping coefficients (m^{-1}).

Wave height (m)	Period (sec)						
	20	30	40	50	60	70	
10	$.36*10^{-4}$	$.33*10^{-4}$	$.30*10^{-4}$	$.27*10^{-4}$	$.25*10^{-4}$	$.24*10^{-4}$	$d = 30$ m
20	$.99*10^{-4}$	$.89*10^{-4}$	$.79*10^{-4}$	$.72*10^{-4}$	$.66*10^{-4}$	$.61*10^{-4}$	
10	$.99*10^{-5}$	$.10*10^{-4}$	$.94*10^{-5}$	$.87*10^{-5}$	$.81*10^{-5}$	$.77*10^{-5}$	
20	$.87*10^{-4}$	$.27*10^{-4}$	$.25*10^{-4}$	$.23*10^{-4}$	$.21*10^{-4}$	$.20*10^{-4}$	$d = 50$ m
30	$.48*10^{-4}$	$.48*10^{-4}$	$.44*10^{-4}$	$.40*10^{-4}$	$.37*10^{-4}$	$.35*10^{-4}$	
40	**	$.72*10^{-4}$	$.66*10^{-4}$	$.60*10^{-4}$	$.55*10^{-4}$	$.52*10^{-4}$	
10	$.39*10^{-5}$	$.45*10^{-5}$	$.43*10^{-5}$	$.41*10^{-5}$	$.38*10^{-5}$	$.36*10^{-5}$	
20	$.10*10^{-4}$	$.12*10^{-4}$	$.11*10^{-4}$	$.11*10^{-4}$	$.10*10^{-4}$	$.94*10^{-5}$	
30	$.19*10^{-4}$	$.21*10^{-4}$	$.20*10^{-4}$	$.19*10^{-4}$	$.17*10^{-4}$	$.16*10^{-4}$	$d = 70$ m
40	$.29*10^{-4}$	$.32*10^{-4}$	$.30*10^{-4}$	$.28*10^{-4}$	$.26*10^{-4}$	$.24*10^{-4}$	
50	**	$.44*10^{-4}$	$.42*10^{-4}$	$.38*10^{-4}$	$.36*10^{-4}$	$.33*10^{-4}$	
10	$.18*10^{-5}$	$.24*10^{-5}$	$.24*10^{-5}$	$.23*10^{-5}$	$.22*10^{-5}$	$.21*10^{-5}$	
20	$.48*10^{-5}$	$.64*10^{-5}$	$.63*10^{-5}$	$.60*10^{-5}$	$.57*10^{-5}$	$.54*10^{-5}$	
30	$.87*10^{-5}$	$.11*10^{-4}$	$.11*10^{-4}$	$.11*10^{-4}$	$.10*10^{-4}$	$.94*10^{-4}$	
40	$.13*10^{-4}$	$.17*10^{-4}$	$.17*10^{-4}$	$.16*10^{-4}$	$.15*10^{-4}$	$.14*10^{-4}$	$d = 90$ m
50	$.18*10^{-4}$	$.24*10^{-4}$	$.23*10^{-4}$	$.22*10^{-4}$	$.20*10^{-4}$	$.19*10^{-4}$	
60	**	$.31*10^{-4}$	$.30*10^{-4}$	$.28*10^{-4}$	$.26*10^{-4}$	$.24*10^{-4}$	
70	**	**	$.37*10^{-4}$	$.35*10^{-4}$	$.32*10^{-4}$	$.30*10^{-4}$	
10	$.88*10^{-6}$	$.14*10^{-5}$	$.15*10^{-5}$	$.14*10^{-5}$	$.14*10^{-5}$	$.13*10^{-5}$	
20	$.24*10^{-5}$	$.30*10^{-5}$	$.39*10^{-5}$	$.38*10^{-5}$	$.36*10^{-5}$	$.34*10^{-5}$	
30	$.44*10^{-5}$	$.68*10^{-5}$	$.69*10^{-5}$	$.66*10^{-5}$	$.63*10^{-5}$	$.60*10^{-5}$	
40	$.66*10^{-5}$	$.10*10^{-4}$	$.10*10^{-4}$	$.99*10^{-4}$	$.94*10^{-5}$	$.89*10^{-5}$	$d = 110$ m
50	$.91*10^{-5}$	$.14*10^{-4}$	$.14*10^{-4}$	$.14*10^{-4}$	$.13*10^{-4}$	$.12*10^{-4}$	
60	$.12*10^{-4}$	$.18*10^{-4}$	$.18*10^{-4}$	$.18*10^{-4}$	$.15*10^{-4}$	$.16*20^{-4}$	
70	**	$.23*10^{-4}$	$.23*10^{-4}$	$.22*10^{-4}$	$.20*10^{-4}$	$.19*10^{-4}$	

** Wave steepness exceeds 0.14 tanh(kh).

Table 6.2. Dissipation by the turbulent boundary layer medium sand (by USC Standard). $D_s = 0.5$ mm, $SG = 2.65$ damping coefficients D_B (m^{-1}).

Wave height (m)	Period (sec)					
	20	30	40	50	60	70
10	$.24 * 10^{-4}$	$.23 * 10^{-4}$	$.20 * 10^{-4}$	$.19 * 10^{-4}$	$.17 * 10^{-4}$	$.16 * 10^{-4}$
20	$.88 * 10^{-4}$	$.80 * 10^{-4}$	$.70 * 10^{-4}$	$.64 * 10^{-4}$	$.58 * 10^{-4}$	$.54 * 10^{-4}$
10	$.50 * 10^{-5}$	$.56 * 10^{-5}$	$.53 * 10^{-5}$	$.49 * 10^{-5}$	$.46 * 10^{-5}$	$.43 * 10^{-5}$
20	$.22 * 10^{-4}$	$.22 * 10^{-4}$	$.20 * 10^{-4}$	$.19 * 10^{-4}$	$.17 * 10^{-4}$	$.16 * 10^{-4}$
30	$.44 * 10^{-4}$	$.44 * 10^{-4}$	$.40 * 10^{-4}$	$.37 * 10^{-4}$	$.37 * 10^{-4}$	$.34 * 10^{-4}$
40	$.32 * 10^{-4}$	$.70 * 10^{-4}$	$.64 * 10^{-4}$	$.58 * 10^{-4}$	$.53 * 10^{-4}$	$.50 * 10^{-4}$
10	$.12 * 10^{-5}$	$.19 * 10^{-5}$	$.20 * 10^{-5}$	$.18 * 10^{-5}$	$.17 * 10^{-5}$	$.16 * 10^{-5}$
20	$.75 * 10^{-5}$	$.91 * 10^{-5}$	$.88 * 10^{-5}$	$839 * 10^{-5}$	$.77 * 10^{-5}$	$.73 * 10^{-5}$
30	$.16 * 10^{-4}$	$.19 * 10^{-4}$	$.18 * 10^{-4}$	$.16 * 10^{-4}$	$.15 * 10^{-4}$	$.14 * 10^{-4}$
40	$.26 * 10^{-4}$	$.30 * 10^{-4}$	$.28 * 10^{-4}$	$.26 * 10^{-4}$	$.24 * 10^{-4}$	$.23 * 10^{-4}$
50	**	$.43 * 10^{-4}$	$.40 * 10^{-4}$	$.37 * 10^{-4}$	$.34 * 10^{-4}$	$.32 * 10^{-4}$
10	$.21 * 10^{-5}$	$.69 * 10^{-6}$	$.79 * 10^{-6}$	$.78 * 10^{-6}$	$.75 * 10^{-6}$	$.70 * 10^{-6}$
20	$.30 * 10^{-5}$	$.45 * 10^{-5}$	$.46 * 10^{-5}$	$.44 * 10^{-5}$	$.42 * 10^{-5}$	$.39 * 10^{-5}$
30	$.68 * 10^{-5}$	$.95 * 10^{-5}$	$.94 * 10^{-5}$	$.90 * 10^{-5}$	$.84 * 10^{-5}$	$.80 * 10^{-5}$
40	$.11 * 10^{-4}$	$.15 * 10^{-4}$	$.15 * 10^{-4}$	$.14 * 10^{-4}$	$.13 * 10^{-4}$	$.13 * 10^{-4}$
50	$.17 * 10^{-4}$	$.22 * 10^{-4}$	$.22 * 10^{-4}$	$.20 * 10^{-4}$	$.19 * 10^{-4}$	$.18 * 10^{-4}$
60	**	$.30 * 10^{-4}$	$.29 * 10^{-4}$	$.27 * 10^{-4}$	$.25 * 10^{-4}$	$.24 * 10^{-4}$
70	**	**	$.37 * 10^{-4}$	$.35 * 10^{-4}$	$.32 * 10^{-4}$	$.30 * 10^{-4}$
10	$.11 * 10^{-5}$	$.14 * 10^{-5}$	$.16 * 10^{-5}$	$.28 * 10^{-6}$	$.27 * 10^{-6}$	$.24 * 10^{-6}$
20	$.12 * 10^{-5}$	$.25 * 10^{-5}$	$.25 * 10^{-5}$	$.26 * 10^{-5}$	$.25 * 10^{-5}$	$.24 * 10^{-5}$
30	$.31 * 10^{-5}$	$.54 * 10^{-5}$	$.56 * 10^{-5}$	$.54 * 10^{-5}$	$.52 * 10^{-5}$	$.49 * 10^{-5}$
40	$.53 * 10^{-5}$	$.39 * 10^{-5}$	$.92 * 10^{-5}$	$.88 * 10^{-5}$	$.84 * 10^{-5}$	$.49 * 10^{-5}$
50	$.79 * 10^{-5}$	$.13 * 10^{-4}$	$.13 * 10^{-4}$	$.13 * 10^{-4}$	$.12 * 10^{-4}$	$.11 * 10^{-4}$
60	$.11 * 10^{-4}$	$.17 * 10^{-4}$	$.18 * 10^{-4}$	$.17 * 10^{-4}$	$.16 * 10^{-4}$	$.15 * 10^{-4}$
70	**	$.22 * 10^{-4}$	$.22 * 10^{-4}$	$.21 * 10^{-4}$	$.20 * 10^{-4}$	$.19 * 10^{-4}$

** Wave steepness exceeds 0.14 tanh(kh).

Table 6.3. Dissipation by the turbulent boundary layer coarse sand (by USC Standard). $D_s = 1.5$ mm, $SG = 2.65$ damping coefficients D_B (m^{-1}).

Wave height (m)	Period (sec)					
	20	30	40	50	60	70
10	$.19*10^{-4}$	$.18*10^{-4}$	$.17*10^{-4}$	$.16*10^{-4}$	$.16*10^{-4}$	$.16*10^{-4}$
20	$.44*10^{-4}$	$.41*10^{-4}$	$.37*10^{-4}$	$.33*10^{-4}$	$.30*10^{-4}$	$.28*10^{-4}$
10	$.12*10^{-5}$	$.11*10^{-4}$	$.99*10^{-5}$	$.90*10^{-5}$	$.83*10^{-5}$	$.78*10^{-5}$
20	$.41*10^{-5}$	$.62*10^{-5}$	$.60*10^{-5}$	$.54*10^{-5}$	$.47*10^{-5}$	$.43*10^{-5}$
30	**	$.24*10^{-4}$	$.26*10^{-4}$	$.24*10^{-4}$	$.22*10^{-4}$	$.20*10^{-4}$
40	**	$.50*10^{-4}$	$.46*10^{-4}$	$.42*10^{-4}$	$.39*10^{-4}$	$.36*10^{-4}$
10	$.53*10^{-5}$	$.52*10^{-5}$	$.50*10^{-5}$	$.47*10^{-5}$	$.43*10^{-5}$	$.40*10^{-5}$
20	$.38*10^{-5}$	$.29*10^{-5}$	$.27*10^{-5}$	$27*10^{-5}$	$.28*10^{-5}$	$.28*10^{-4}$
30	$.58*10^{-5}$	$.87*10^{-5}$	$.87*10^{-5}$	$.82*10^{-5}$	$.76*10^{-5}$	$.71*10^{-5}$
40	$.15*10^{-4}$	$.19*10^{-4}$	$.18*10^{-4}$	$.17*10^{-4}$	$.16*10^{-4}$	$.15*10^{-4}$
50	**	$.31*10^{-4}$	$.30*10^{-4}$	$.28*10^{-4}$	$.25*10^{-4}$	$.24*10^{-4}$
10	$.50*10^{-5}$	$.30*10^{-5}$	$.27*10^{-5}$	$.26*10^{-5}$	$.24*10^{-5}$	$.23*10^{-5}$
20	$.53*10^{-5}$	$.30*10^{-5}$	$.28*10^{-5}$	$.27*10^{-5}$	$.26*10^{-5}$	$.26*10^{-5}$
30	$.76*10^{16}$	$.31*10^{-5}$	$.35*10^{-5}$	$.35*10^{-5}$	$.33*10^{-5}$	$.31*10^{-5}$
40	$.48*10^{-5}$	$.85*10^{-5}$	$.88*10^{-5}$	$.85*10^{-5}$	$.80*10^{-5}$	$.75*10^{-5}$
50	**	$.94*10^{-5}$	$.15*10^{-4}$	$.15*10^{-4}$	$.14*10^{-4}$	$.13*10^{-4}$
60	**	$.12*10^{-4}$	$.22*10^{-4}$	$.22*10^{-4}$	$.21*10^{-4}$	$.19*10^{-4}$
70	**	**	$.30*10^{-4}$	$.28*10^{-4}$	$.26*10^{-4}$	$.24*10^{-4}$
10	$.50*10^{-5}$	$.23*10^{-5}$	$.19*10^{-5}$	$.19*10^{-5}$	$.16*10^{-5}$	$.14*10^{-5}$
20	$.30*10^{-5}$	$.25*10^{-5}$	$.24*10^{-5}$	$.23*10^{-5}$	$.22*10^{-5}$	$.22*10^{-5}$
30	$.17*10^{-5}$	$.66*10^{-5}$	$.13*10^{-5}$	$.14*10^{-5}$	$.13*10^{-5}$	$.12*10^{-5}$
40	$.12*10^{-5}$	$.41*10^{-5}$	$.46*10^{-5}$	$.46*10^{-5}$	$.44*10^{-5}$	$.42*10^{-5}$
50	$.34*10^{-5}$	$.77*10^{-5}$	$.83*10^{-5}$	$.81*10^{-5}$	$.77*10^{-5}$	$.73*10^{-5}$
60	$.60*10^{-5}$	$.12*10^{-4}$	$.12*10^{-4}$	$.12*10^{-4}$	$.11*10^{-4}$	$.11*10^{-4}$
70	**	$.16*10^{-4}$	$.17*10^{-4}$	$.16*10^{-4}$	$.15*10^{-4}$	$.15*10^{-4}$

** Wave steepness exceeds 0.14 $\tanh(kh)$.

Table 6.4. Computed friction factors for wind-generated waves. Bed material is coarse sand (by UCS Standard). $D_s = 1.5$ mm, $SG = 2.65$.

Wave height (m)	Period (sec)		
	8	12	16
4	.077	.058	.046 $d = 10$ m
8	**	**	.010
4	.210	.127	.062
8	.078	.053	.020 $d = 20$ m
12	**	.008	.010
4	.307	.169	.138
8	.199	.056	.046
12	.092	.022	.013 $d = 30$ m
16	**	.004	.011
20	**	**	.023
4	.349	.196	.151
8	.287	.086	.044
12	.222	.058	.045 $d = 40$ m
16	**	.021	.008
20	**	.007	.012
4	.605	.228	.164
8	.324	.157	.055
12	.225	.059	.046 $d = 50$ m
16	**	.058	.018
20	**	.055	.006

** Wave steepness exceeds 0.14 tanh(kh).

6. Response to Wave-Induced Pressure Fluctuations

In addition to the dissipation of wave energy by the oscillating, turbulent bottom boundary layer, a significant amount of wave energy may be dissipated within the seabed by various relaxation mechanisms. This is particularly important on oozy seafloors such as those encountered near the estuary of the Mississippi River. Therefore, it may also be important for EGWWs at some locations. Unfortunately, at this time, this effect has not been assessed quantitatively for EGWWs.

Table 6.5. Friction factors as computed from field experiments.

Reference	Location	Friction factor	Range	Remarks
Bretschneider (1954)	Gulf of Mexico	$f_w = 0.106$	0.060–1.934	
Bretschneider (1954)	Gulf of Mexico	$f_w = 0.02$		
	Akita Coast	$f_w = 0.116$	0.066–0.180	
	Izumisano Coast	$f_w = 0.280$		
	Hiezu Coast (1963)	$f_w = 0.166$	0.054–0.260	
Iwagaki and Kakinuma (1967)	Nishikinohana Coast	$f_w = 1.100$	0.560–2.320	20–40 ft. slope: 0.006
	Hienzu Coast (1964)	$f_w = 0.094$	0.020–0.140	
	Takehama Coast	$f_w = 0.100$	0.060–0.160	
Kishi (1975)	Nirigata	$f_w = 0.035$	0.03–0.04	6–8 ft. slope: 0.018
Hasselman Collins (1968)	Gulf of Mexico	$f' = 0.015$		Hurricane Hilda
Van Leperen (1975)	Melkbosstrand	$f' =$	0.06–0.10	
Breeding (1972)	Gulf of Mexico	$f' =$	0.035–0.05	
Shemdin et al. (1975)	Florida Coast	$f' = 0.008$	0.002–0.05	
Collins (1972)		$f' = 0.015$		Hurricane Dora, Betsy

Note: $f' = \frac{1}{2}, f_w = f$.

The subsoil characteristics (which are site specific) are less well known when the sediment surface layer and the subsoil are highly stratified. Most of the existing theory deals only with homogeneous soil.

Of paramount importance in the prediction of the damping of EGWWs by soil response is the knowledge of the subsoil characteristics. This *in situ* information remains one of the main difficulties for the application of the theoretical developments. It is, for the time being, an area of uncertainty.

The interactions that occur between surface gravity waves and the seabed are modelled most generally by treating the seabed as a porous, elastic (hence the term "poroelastic") medium. Past investigators have, however, treated the bed as a heavy viscous fluid (Gade, 1958), or as a rigid permeable medium through which only percolation is allowed (Putnam, 1949; Reid and Kajiura, 1957; Liu, 1973; Liu and Dalrymple, 1984). The works of Biot (1956 and 1962) are now heavily relied on in the establishment of the constitutive relations and the equations of motion for the marine sediment. A semi-analytic solution of these equations of motion allows us to derive a coupled, modified dispersion relation for surface gravity waves propagating over such seabeds from which the rate of damping is determined.

Specifically, the seabed is treated as an isotropic, homogeneous layer of finite thickness which is assumed to be underlain by a halfspace of impervious, infinitely rigid bedrock. Previous work treating the bed as a poroelastic medium analyzed the interaction between an infinite, homogeneous halfspace (Yamamoto, 1983). In addition, inhomogeneities and anisotropy of the seabed have been taken into account utilizing the propagator matrix method to analyze the propagation of acoustic waves in the seabed (Yamamoto, 1983).

An attempt in applying the theory of Yamamoto to the case of finite sediment depth has been developed by Swan (1984). Applications to EGWWs have not been done at the time of this writing. Therefore, the subject here is only mentioned for future developments in the state of the art of EGWWs.

7. Dissipation by Wave Breaking

As previously indicated, EGWWs have often been compared to tsunami waves. From a dynamic point of view, the differences are very significant particularly when the wave reaches the gentle slopes encountered near the coastlines. A major difference is that the tsunami wave does not break before it reaches the shore, whereas EGWWs of interest always breaks offshore. An

EGWW of large amplitude may even break at the edge of the continental shelf.

To determine whether a wave breaks on a slope, it suffices to verify that the wave steepness is larger than (Miche, 1944),

$$\frac{H}{L}\bigg|_{max} > \left(\frac{2\alpha}{\pi}\right)^{1/2} \frac{\sin^2 \alpha}{\pi}, \tag{6.36}$$

where α is the slope. Equation (6.36) indicates clearly that a very long wave ($\frac{H}{L}$ small) does not break even on relatively gentle slopes. The waves run up the shore and reflect seawards. At times a tsunami wave can form a tidal bore nearshore, characterized by a near vertical wall of water, which is very different from the breakers forming at the crests of EGWWs. The breaking of EGWWs is also very different from the breaking of wind waves (swell or sea). Indeed, if one assumes that EGWWs have a very large amplitude, their breaking inception is further offshore than wind waves. In general, it means that EGWWs break on a very gentle slope (say less than 1/100) such as those encountered on continental shelves, whereas wind waves break on the steeper beach slope (say 1/10 to 1/50).

The consequences of this are important to determine the wave run-up as it has been explained in Chapter 1: EGWWs form a nonsaturated breaker, the height of which is depth controlled, up to the point where the beach slope steepens (Le Méhauté, 1963; Divoky *et al.*, 1970). We will not further extend on this subject which has been well covered in the past.

Chapter 7

PROPAGATION OF TRANSIENT WAVES ON NONUNIFORM BATHYMETRIES

1. Basic Principles — A Review

Studies on the transformation of transient waves at nonuniform water depths are limited to the case of linear waves. The usual approach is to follow the transport of wave energy. This has been employed by a number of researchers interested in the propagation of EGWWs. Van Dorn (1964) gave a formula for calculating the effect of frequency dispersion of a wave normally incident on a plane beach. Van Mater and Neal (1970) employs Van Dorn's (1964) frequency dispersion factor in an earlier attempt to model the behavior of impulsively generated waves as they propagate in shallow water. Le Méhauté (1971) described a somewhat more general method in which ray separation for periodic waves together with Van Dorn's frequency dispersion factor are used to propagate explosion waves. Houston and Chou (1983) used a combination of radial spreading and plane wave refraction to propagate explosion waves from the edge of continental shelf into shallow water. All these theories are approximate and in some way empirical.

Now the state of the art has progressed to the point where exact methods can be used and applied to linear transient·waves. There are essentially three exact methods. None of which are simple. The level of difficulty varies with the intricacies of the arithmetic and the corresponding numerical and computer programs.

The first method is based on the conservation of wave energy spectrum in terms of wave numbers k, as it is used in the study of the propagation of random irregular waves. Given the time history of an EGWW at a given point and the corresponding wave directions, the wave energy spectrum is determined by standard methods using FFT. The directional spectrum at that point is obtained. In the case where the point is in deep water, and the spectrum is defined in terms of frequency σ and direction θ_0 by $S_0(\sigma, \theta_0)$, then the shallow water spectrum is given simply by (Pierson, Tuttell and Woodley, 1953):

$$S(\sigma, \theta) = S_0(\sigma, \theta_0) K_R^2 K_S^2 \frac{\partial \theta}{\partial \theta_0} \tag{7.1}$$

The shallow water wave direction, θ, is obtained by refraction as a function of σ and θ_0. K_R and K_S are the refraction and shoaling coefficients respectively. In the case of the crossing of wave orthogonals (caustic): $K_R \to \infty$, $\frac{\partial \theta}{\partial \theta_0} - \infty$, but their product is finite as demonstrated by Le Méhauté and Wang (1980). It is found that:

$$K_R^2 \frac{\partial \theta}{\partial \theta_0} = \frac{k}{k_0}, \quad \text{and} \quad K_S^2 = \frac{V_0}{V}, \tag{7.2}$$

which is always finite. Since K_S^2 is the ratio of group velocity (V_0/V), the transform operator is simply: (kV_0/k_0V).

This method is relatively simple, but unlike random waves, one is now concerned with the time history and phase of EGWWs, information which is lost in the energy spectrum. Therefore, another equation is necessary. In theory, since the energy as a function of wave number is well determined and conserved, it suffices to follow the wave elements with constant wave numbers, to calculate its phase φ along the wave rays:

$$\varphi(t) = \varphi_{t=t_0} + \sigma(t - t_0) - \int_{t_0}^t V(k)k\,dt. \tag{7.3}$$

Therefore, in theory, the time history of EGWWs is obtained by combining this information with an inverse FFT of the shallow water spectrum. In practice, this method is very complex.

An alternate, simpler approach to the study of linear transient waves consists of decomposing a wave at a given location into components of varying frequency σ. Each component is propagated into a fixed, shallow-water location $\bar{x}$. The wave train is then reconstructed using an inverse Fourier integral of the form

$$\eta(\bar{x}, t) = \int_0^\infty A(\bar{x}, \sigma)e^{i\sigma t}d\sigma, \tag{7.4}$$

where $A(\bar{x}, \sigma)$ is the complex amplitude of the component at frequency σ. This amplitude includes the effect of linear refraction, shoaling, and radial dispersion. This transformation is in the frequency domain, which requires initial conditions at a given point as a function of time.

Since the initial conditions are often given at time zero as a function of distance, the amplitude spectrum is obtained in terms of wave number instead of frequency. Therefore, a number of transformations needs to be done initially. This method allows some convenient approximations, such as the use of asymptotic solutions, which are valid at a distance from GZ, and make the problem tractable. It is the most convenient method in the case of complex 3D bathymetry which is detailed in the following.

For accuracy, one has to rely on the principle of conservation of wave action between wave rays applied to a wave patch traveling at group velocity as shown in Figure 7.1 (Le Méhauté and Soldate, 1986; Soldate *et al.*, 1986). Since the limit of the patch AB and CD have different frequencies, they travel at different velocities, along different wave rays, so that the patch shape and area vary continuously.

This method, even though based on the linear appoximation, allows the introduction of correction factors due to nonlinear effects. These include dissipative processes such as wave-seafloor interaction as presented in Chapter 6 and convective effects on wave shoaling as indicated in Chapter 1. The method has been applied to the propagation of EGWWs from deep to shallow water. It is extremely complex; the corresponding numerical scheme is resolved step by step using twelve simultaneous differential equations. Also, the problem of transient caustics, which occurs over 3D bathymetries has not been resolved. Because of its limitations and complexity, the method has been applied to plane bathymetries only. Its advantage is the possibility of taking into account some nonlinear effects, such as energy dissipation, which are important over a long shelf. Figures 1.23–1.31 have been obtained by the application of this method. This method can be found in detail in Le Méhauté and Soldate (1986) and are not presented here.

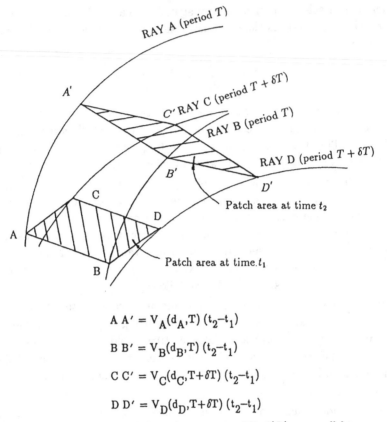

$$AA' = V_A(d_A, T)\,(t_2 - t_1)$$

$$BB' = V_B(d_B, T)\,(t_2 - t_1)$$

$$CC' = V_C(d_C, T + \delta T)\,(t_2 - t_1)$$

$$DD' = V_D(d_D, T + \delta T)\,(t_2 - t_1)$$

Fig. 7.1. Evolution of a wave patch from time t_1 to t_2. AB, $A'B'$ are parallel to wave crests at period T and CD, $C'D'$ are parallel to wave crests at period $T + \delta T$.

In the following, the method based on Fourier integral technique is presented for a 3D bathymetry and verified experimentally.

2. Evolution of a Periodic Wave from a Point Source over a 3D Bathymetry

Consider a periodic wave of frequency σ, at a point source $T_0(x_0, y_0)$ or $T_0(r_0, \theta_0)$ in the form of a delta function of amplitude $A_0(r_0, \theta_0)$. In the case of a horizontal seafloor, the wave motion at any point $T(x, y)$ can be represented by

$$\eta(x,\, y,\, t) = A_0(r_0,\, \theta_0) J_0(k\bar{r}) \cos \sigma t\,, \tag{7.5}$$

where $\bar{r}$ is the distance between $T_0(x_0,\, y_0)$ and $T(x,\, y)$. When $(k\bar{r})$ is large, using the asymptotic form of the Bessel function J_0, this equation can be approximated by

$$\eta(x,\, y,\, t) = A_0(r_0,\, \theta_0) \left[\frac{2}{\pi k \bar{r}} \right]^{\frac{1}{2}} \cos\left(k\bar{r} - \frac{\pi}{4} \right) \cos \sigma t\,. \tag{7.6}$$

When the seafloor is no longer horizontal but defined by a three-dimensional bathymetry, then:

$$\eta(x,\, y,\, t) = A_0(r_0,\, \theta_0) |K_R K_S K_T|_{r_0, \theta_0}^{x, y} \cos \sigma t\,, \tag{7.7}$$

where K_R, K_S are the refraction and shoaling coefficients, respectively, for parallel, long crested waves traveling along the wave ray s between $T_0(r_0,\, \theta_0)$ and $T(x,\, y)$, at frequency σ. K_T is the angular dispersion coefficient which, for large values of $\bar{r}$, can be approximated in amplitude and phase as:

$$K_T(\sigma) \cong \left[\frac{2}{\pi k s} \right]^{1/2} \cos\left(\int_{r_0, \theta_0}^{s} k \, ds - \frac{\pi}{4} \right). \tag{7.8}$$

Therefore,

$$\eta(x,\, y,\, t) \cong A_0(r_0,\, \theta_0) |K_R(\sigma) K_S(\sigma)|_{r_0, \theta_0}^{s} \left[\frac{2}{\pi k s} \right]^{1/2} \cos\left(\int_0^s k \, ds - \frac{\pi}{4} \right) \cos \sigma t\,. \tag{7.9}$$

The shoaling coefficient K_S is

$$K_S = \left\{ \frac{\tanh k \left[1 + \dfrac{2k}{\sinh 2k} \right]}{\tanh k_T \left[1 + \dfrac{2k_T}{\sinh 2k_T} \right]} \right\}^{1/2}. \tag{7.10}$$

where k is the dimensionless wave number ($k = k^* d$) at the source ($r_0,\, \theta_0$) where the water depth is d and k_T is the dimensionless wave number ($k_T = k_T^* d_T$) at $T(x,\, y)$ where the water depth is d_T.

The determination of K_R is in general much more complex except for parallel bottom contours. In this case, let θ and θ_T be the angle between a wave ray and a line perpendicular to the bottom contours at the origin and at T, respectively. Then by virtue of the principle of conservation of energy between wave orthogonals:

$$K_R = \left| \frac{\cos \theta}{\cos \theta_T} \right|^{1/2} . \tag{7.11}$$

Applying Snell's law,

$$k^* \sin \theta = k_T^* \sin \theta_T , \tag{7.12}$$

$$K_R = \left\{ \frac{\cos \left[\sin^{-1} \left(\frac{k_T^*}{k^*} \sin \theta_T \right) \right]}{\cos \theta_T} \right\}^{1/2}$$

$$= \left\{ \frac{\cos \theta}{\cos \left[\sin^{-1} \left(\frac{k_T^*}{k^*} \sin \theta_T \right) \right]} \right\}^{1/2} . \tag{7.13}$$

In general, there is only one possible path s for each frequency which ensures that the wave ray goes from $T(r_0, \theta_0)$ to the considered target $T(x, y)$ and vice versa, there is only one value for a reverse way ray to go from T_0 to T. However, if an island or a mound interfers between T_0 and T, there could be two paths for each frequency.

In general, the wave rays have to be determined by trial and error for each frequency. In many practical cases, taking s along a straight line between the two points is a sufficient approximation which reduces the calculation by orders of magnitude. Further, assuming θ_T as the angle between this pseudo ray and the perpendicular to the average bottom contour at T and applying Equation (7.11) for K_R, permits us to resolve the problem straightforwardly.

When these simplifications are not valid, the refraction coefficient K_R has to be determined from the "ray separation factor" (Munk and Arthur, 1951). Then

$$K_R^2 = \frac{1}{\beta} . \tag{7.14}$$

The determination of β is a standard procedure in the field of water waves, which will not be detailed here.

3. Transformation of a Finite Source from the Wave Number Domain to the Frequency Domain

Now consider the case of a axisymmetrical source of finite dimensions defined by a free-surface elevation $\eta_0(r_0, \theta_0)$ at $t = 0$ such as $\partial\eta/\partial\theta = 0$, $0 \leq r_0 \leq R$, and $\partial\eta_0/\partial t = 0$. The seafloor is assumed to be horizontal over the area source, and becomes three-dimensional when $r > R$.

Recall that the amplitude spectrum at time $t = 0$ in terms of wave number is given by the double Fourier integral

$$B(k) = \frac{1}{2\pi} \int_{-\infty}^{\infty} \int_{-\infty}^{\infty} \eta_0(x, y) \, \exp^{-i(k_x x + k_y y)} \, dx dy \,, \qquad (7.15)$$

where $\bar{k}(k_x, k_y)$ is the wave number vector. Note the relationship

$$k_x x + k_y y = kr \cos(\theta_0 - \phi) \,, \qquad (7.16)$$

where $(\theta_0 - \phi)$ is the phase angle between the vectors $\bar{k}$ and $\bar{r}$. Because of the symmetry at the origin, the above relationship is independent of the value of ϕ, which is then taken equal to zero. Considering further,

$$\eta(x, y) = \eta(r_0, \theta_0) = \eta(r_0) \,, \qquad (7.17)$$

the above relationship can be written:

$$B_0(k) = \frac{1}{2\pi} \int_{-\infty}^{2\pi} \int_{-\infty}^{\infty} \eta(r_0) \, \exp[-ikr_0 \cos \theta_0] r_0 dr_0 d\theta_0 \,. \qquad (7.18)$$

Inserting

$$J_0(kr) = \frac{1}{2\pi} \int_{0}^{2\pi} \exp(-ikr \cos \theta) d\theta \qquad (7.19)$$

yields

$$B_0(k) = \int_{0}^{\infty} \eta_0(r_0) J_0(kr_0) r_0 dr_0 \,, \qquad (7.20)$$

which is the Hankel Transform of the initial free-surface deformation as described in Chapter 2.

Since this amplitude spectrum originates at various locations of the initial disturbance, its transformation with distance over a 3D bathymetry is not trivial. However, it allows the solution of the transient wave at the source, as has been seen previously in Chapter 2. At the source, the wave motion is described by the general solution over a horizontal seafloor for a value of $r = r_0$. Then:

$$\eta_0(r_0, \theta_0, t) = \int_0^\infty B_0(k) J_0(kr_0) \cos \sigma t \, k \, dk . \tag{7.21}$$

Each of these points (r_0, θ_0) originate an elementary transient wave over the 3D bathymetry outside the source. This transformation is now analyzed.

When the wave motion at $T_0(r_0, \theta_0)$ is not periodic, but defined by a transient $\eta_0(r_0, \theta_0, t)$, as given by Equation (7.21), the wave motion at this location can be considered as the sum of an infinite number of periodic wave of amplitude $A_{r_0}, \theta_0(\sigma)$ as a function of frequency σ.

The amplitude spectrum in terms of frequency which characterizes the wave motion at (r_0, θ_0) is then given by the Fourier Transform:

$$A_{r_0}, \theta_0(\sigma) = \int_{-\infty}^\infty \eta_0(r_0, \theta_0, t) e^{i\sigma t} dt . \tag{7.22}$$

The amplitude spectrum at a distance $T(x, y)$ from (r_0, θ_0) is given by multiplying all the values of $A_{r_0}, \theta_0(\sigma)$ by the transformation coefficients $K_R(\sigma)$, $K_S(\sigma)$, $K_T(\sigma)$ along the wave rays $s(\sigma)$ between $T_0(r_0, \theta_0)$ and $T(x, y)$. Note that the transformation coefficients as well as the wave rays $s(\sigma)$ have to be determined for all the frequencies. Then the amplitude spectrum at $T(x, y)$ resulting from $A_{r_0,\theta_0}(\sigma)$ is:

$$A_{x,y}(\sigma) = A_{r_0,\theta_0}(\sigma) |K_R(\sigma) K_S(\sigma) K_T(\sigma)|_{r_0,\theta_0}^{x,y} . \tag{7.23}$$

The time history of the free-surface elevation at $T(x, y)$ due to a source location (r_0, θ_0) is given by the reverse Fourier Transform:

$$\eta_{r_0,\theta_0}(x, y, t) = \int_{-\infty}^\infty A_{x,y}(\sigma) e^{-i\sigma t} d\sigma . \tag{7.24}$$

The total wave motion at $T(x, y)$ is given by the sum of all the transient wave components $\eta_{r0,\theta0}(x, y, t)$ originating at each point (r_0, θ_0) over the total area source, such as

$$\eta(x, y, t) = \frac{1}{\pi R^2} \int_0^{2\pi} \int_0^R \eta_{r_0,\theta_0}(x, y, t) r_0 dr_0 d\theta_0 . \tag{7.25}$$

As seen, the determination of $\eta(x, y, t)$ requires a large number of integrals which are summarized in Table 7.1. The most demanding part is to determine

Table 7.1. A summary of equations showing the wave spectral contents and transformation.

Initial condition	$\eta_0(r_0\,\theta_0) = \eta_0(r_0)$
Amplitude spectrum in the wave number domain	$B_0(k) = \displaystyle\int_0^{R_0} \eta_0(r_0) J_0(kr_0) r_0 dr_0$
Transient wave motion at the source $0 < r = r_0 \leq R_0$	$\eta_0(r_0,\,\theta_0,\,t) = \displaystyle\int_0^\infty B_0(k) J_0(kr_0)\cos\sigma t\, k\, dk$
Amplitude spectrum in the frequency domain at the source $(r_0,\,\theta_0)$	$A_{r_0,\theta_0}(\sigma) = \displaystyle\int_{-\infty}^\infty \eta_0(r_0,\,\theta_0,\,t) e^{-i\sigma t} dt$
Amplitude spectrum in the frequency domain at $T(x,\,y)$ due to a point source $T_0(r_0,\,\theta_0)$	$A_{x,y}(\sigma) = A_{r_0,\theta_0}(\sigma)\|K_S(\sigma)K_R(\sigma)K_T(\sigma)\|_{r_0,\theta_0}^{x,y}$
Wave motion at $T(x,\,y)$ due to a point source $(r_0,\,\theta_0)$	$\eta_{r_0,\theta_0}(x,\,y,\,t) = \dfrac{1}{2\pi}\displaystyle\int_{-\infty}^\infty A_{x,y}(\sigma) e^{i\sigma t} d\sigma$
Wave motion at $T(x,\,y)$ due to the all area source $(0 \leq r_0 < R)$	$\eta(x,\,y,\,t) = \dfrac{1}{\pi R^2}\displaystyle\int_0^{2\pi}\int_0^R \eta_{r_0,\theta_0}(x,\,y,\,t) r_0 dr_0\, d\theta_0$

the wave rays $s(\sigma)$ for all frequencies and from all points $T_0(r_0,\,\theta_0)$ to $T(x,\,y)$, and the values of the corresponding transformation coefficient K_R, K_T (the shoaling coefficient being a function of frequency and water depth only is the same for all points $(r_0,\,\theta_0)$.

It is interesting to show that the above formulation yields previous results for a horizontal seafloor. For a horizontal seafloor, the wave motion is symmetrical about $r = 0$, so that the integral with respect to θ_0 vanishes. Also,

$$K_R K_S = 1 \quad \text{and} \quad K_T = J_0(k\bar{r})\,. \tag{7.26}$$

Considering that A_{xy} is a direct function of k (instead of σ), K_T can be extracted from the integral in σ (inverse Fourier integral). Then the two Fourier integrals cancel each other, so that

$$\eta(x,\, y,\, t) = \frac{1}{\pi R^2} \int_0^{2\pi} \int_0^R \left[\int_0^\infty B_0(k) J_0(kr_0) \cos \sigma t \, J_0(k\bar{r}) k dk \right] r_0 dr_0 d\theta_0 \, . \tag{7.27}$$

According to Graf's addition theorem

$$J_0(kr) = J_0(kr_0) J_0(k\bar{r}) + 2 \sum_{n \geq 1} J_n(kr_0) J_n(k\bar{r}) \cos n(\theta_0 - \theta) \tag{7.28}$$

and integration with respect to θ from 0 to 2π of the summation is zero. Therefore, the product of the two Bessel functions in r and $\bar{r}$ can be replaced by a unique Bessel function in r. The integrand,

$$I = B_0(k) J_0(kr) \cos \sigma T \, , \tag{7.29}$$

then becomes independent of r_0, and the integral yields:

$$\eta(x,\, y,\, t) = \eta(r,\, t) = \int_0^\infty B_0(k) J_0(kr) \cos \sigma t \, k dk \, , \tag{7.30}$$

as previously demonstrated.

4. Analytical Verifications on Area Source Solutions

Following step by step the operations presented in Table 7.1, consider a free-surface deformation at time $t = 0$ such as

$$\eta_0(r_0) = \frac{AR^3}{(R^2 + r_0^2)^{3/2}} \, . \tag{7.31}$$

The amplitude spectrum in terms of wave number is

$$B(k) = \int_0^\infty \frac{AR^3}{(R^2 + r_0^2)^{3/2}} J_0(kr_0) r_0 dr_0 \, , \tag{7.32}$$

which is:

$$B(k) = AR^2 \exp(-kR) \, . \tag{7.33}$$

Accordingly, the general solution is

$$\eta(r,\, t) = \int_0^\infty AR^2 \exp(-kR) J_0(kr) \cos \sigma t\, kdk\,. \tag{7.34}$$

Now consider this solution at $r = 0$ and assuming $\sigma = k$. (It is recalled that one has demonstrated in Chapter 2 that dispersion near the origin is negligible.) Then,

$$\eta(o,\, t) = \int_0^\infty AR^2 \exp(-kR) \cos kt\, kdk \tag{7.35}$$

which can be integrated to give

$$\eta(o,\, t) = A\, \frac{1 - \tau^2}{(1 + \tau^2)^2} \tag{7.36}$$

where

$$t = R\tau\,. \tag{7.37}$$

The amplitude spectrum at $r = o$ in terms of frequency is Equation (7.22)

$$A_0(\sigma) = \frac{1}{2\pi} \int_{-\infty}^\infty \frac{1 - \tau^2}{(1 + \tau^2)^2} [\cos(-i\sigma R\tau) - i\, \sin(-i\sigma R\tau)] d\tau \tag{7.38}$$

The second integral above (sin-term) is odd and equals zero. The first integral (cos-term) is even and equals to

$$A_0(\sigma) = AR^2 \sigma\, \exp(-\sigma R)\,, \tag{7.39}$$

which is the amplitude spectrum in terms of frequency at $r = 0$. Therefore, by reverse transform:

$$\eta(o,\, t) = AR^2 \int_0^\infty \sigma \exp(-\sigma R) \cos \sigma t\, d\sigma\,. \tag{7.40}$$

Introducing the transform operator K_T in the case of a horizontal seafloor yields

$$\eta(r,\, t) = AR^2 \int_0^\infty \sigma \exp(-\sigma R) J_0(kr) \cos \sigma t\, d\sigma\,, \tag{7.41}$$

which is consistent with the general solution (7.39) since dispersion is negligible at the origin ($\sigma = k$) as seen in Chapter 2. This is also compatible with

$$\int_0^\infty A(\sigma)d\sigma = \int_0^\infty B(k)k\,dk\,, \qquad (7.42)$$

i.e.,

$$A(\sigma) = B(k)\frac{k}{V}\,, \qquad (7.43)$$

since the dimensionless group velocity

$$V = \frac{C_g}{(gd)^{1/2}} = 1\,, \qquad (7.44)$$

at the origin.

5. Simplified Formulation

Replacing dk by $d\sigma/V$ in Equation (7.30) allows us to work in the frequency domain directly. Therefore, the transform coefficients $K_S K_R K_T$ are applicable to each frequency under the initial integral. Then the inverse Fourier Transform in σ and the Fourier Transform in t cancel so that one is left with

$$\eta(x, y, t) = \int_0^\infty B_0(k(\sigma))|K_R(s)K_S(\sigma)K_T(\sigma)|_{r_0,\theta_0}^{x,y} \cos \sigma t\, k(\sigma)\, \frac{d\sigma}{V} \qquad (7.45)$$

which is equivalent to transforming the wave amplitude spectrum in the wave number domain directly as follows:

$$\eta(x, y, t) = \int_0^\infty B_1(k)|K_R(k)K_S(k)K_T(k)|_{r_0,\theta_0}^{x,y} \cos \sigma(k) \cos \sigma(k)tk\,dk\,. \qquad (7.46)$$

The previous method can further be simplified when r is large compared to the size of the area source R ($\bar{r} \gg R$). Then, the transformation coefficients K_R, K_T can all be considered identical for all the points (r_0, θ_0) and equal to their values for $r_0 = 0$. Then,

$$K_R K_S K_T|_{r_0,\theta_0}^{x,y} \cong K_R K_S K_T|_0^{x,y}\,. \qquad (7.47)$$

Since these K_i values are no longer functions of r_0 and θ_0, the integrals (7.27) in r_0 and θ_0 are simplified. If one considers further that the K_i coefficients are expressed in terms of wave number k (instead of frequency), then they can be extracted from the integral in σ.

In analogy with the case of a wave on a horizontal seafloor (7.28)

$$K_T|_{r_0,\theta_0}^{x,y} \cong K_T|_0^{x,y} = J_0(kr)\,. \qquad (7.48)$$

For practical use, the wave motion near the origin can then be determined by taking $K_T \cong J_0(kr)$ and calculating K_S and K_R between the origin and $T(x, y)$. Accordingly,

$$\eta(x, y, t) \cong \int_0^\infty B_0(k) J_0(kr) K_S(k) K_R(k) \cos \sigma t k \, dk . \qquad (7.49)$$

Far from the origin, using the asymptotic form for K_T, one obtains:

$$\eta(x, y, t) = \int_0^\infty B_0(k) K_R(k) K_S(k) \left[\frac{2}{\pi k s(\sigma)}\right]^{1/2}$$

$$\times \cos \sigma t \cos \left[\int_0^{s(\sigma)} k dS - \frac{\pi}{4}\right] k dk . \qquad (7.50)$$

The solutions (7.49) and (7.50) can also be generalized to any form of initial disturbances as described in Chapter 2, whether the original disturbance is defined by a free-surface elevation, free-surface velocity, or an impulse. Then the general solutions become

$$\left.\begin{matrix} \eta_c \\ \eta_v \\ \eta_I \end{matrix}\right|(r, t) = \int_0^\infty B_0(k) F(\sigma) \cos\left[-\int_0^{s(\sigma)} k ds + \frac{\pi}{4}\right] \left|\begin{matrix} \cos \sigma t \\ \sin \sigma t \\ \sigma \\ \sigma \sin \sigma t \end{matrix}\right| k \, d\sigma \quad (7.51)$$

where

$$F(\sigma) = K_S(\sigma) K_R(\sigma) \left[\frac{1}{\pi k s(\sigma)}\right]^{1/2} \frac{1}{V(k(\sigma))} , \qquad (7.52)$$

and $B_0(k)$ is any of the Hankel transforms (r_0) as presented in Table 2.1. Equation (7.51) can be integrated by FFT. More simply, all the solutions which have been obtained by making use of the stationary phase approximation (kr large) on a horizontal seafloor, can now be easily modified for their generalization over a 3D bathymetry. For example, considering the case of a crater in the form of a parabola with lip (case 3, Table 2.1), one finds

$$\eta(x, y, t) = \frac{\eta_{o\max} R}{ks(\sigma)} \left(\frac{k V(k)}{-\dfrac{dV}{dk}}\right)^{1/2} |K_R(\sigma) K_S(\sigma)|_0^{s(\sigma)}$$

$$\times J_3(kR) \cos \left(\sigma t - \int_0^{s(\sigma)} k ds\right) \qquad (7.53)$$

6. Propagation of EGWWs on a Gentle Slope

Now consider a wave propagating towards the shoreline along a line perpendicular to the bottom as shown in Figure 7.2. Since the wave crest along that line remains parallel to the bottom contour, it suffices to modify the amplitude of J_0 in Equation (7.21) for shoaling and (converging) refraction, and to change the phase kr by $\int_0^\infty k(r)dr$, k being a function of r through depth variation. Therefore, $\eta(r, t)$ along this line can be written as

$$\eta(r, t) = \int_0^\infty B_0(k)J_0' \cos \sigma t \, kdk \,, \tag{7.54}$$

where

$$J_0' \cong K_S K_R J_0\left(\int_0^r k(r)\, dr \right) \tag{7.55}$$

Referring to Figure 7.2, the slope $m = \frac{h^* - h_r^*}{r^*}$. The effect of shoaling at GZ with respect to infinite depth (H_∞) is:

$$\frac{H_{GZ}}{H_\infty} = \left[\frac{1}{\tanh(k)\left(1 + \dfrac{2k}{\sinh(2k)}\right)} \right]^{1/2} . \tag{7.56}$$

the effect of shoaling at r with respect to H_∞ is:

$$\frac{H_r}{H_\infty} = \left[\frac{1}{\tanh[k_r(1 - Sr)] \cdot \left(1 + \dfrac{2k_r(1 - Sr)}{\sinh(2k_r(1 - s_r))}\right)} \right]^{1/2} . \tag{7.57}$$

Therefore, the net effect of shoaling at r with respect to a wave at GZ is

$$\frac{H_r}{H_{GZ}} = K_S = \left\{ \frac{\tanh(k)\left[1 + \dfrac{2k}{\sinh(2k)}\right]}{\tanh[k_r(1 - Sr)] \cdot \left(1 + \dfrac{2k_r(1 - Sr)}{\sinh(2k_r(1 - S_r))}\right)} \right\}^{1/2} . \tag{7.58}$$

From Figure 7.2, the refraction coefficient is

$$K_R = \left(\frac{EF}{EF}\right)^{1/2} = \left[\frac{r \tan(\alpha)}{\Sigma r_A \tan(\alpha_0) + (r_C - r_A)\tan(\alpha_1) + (r_E - r_C)\tan(\alpha_2)} \right]^{1/2} , \tag{7.59}$$

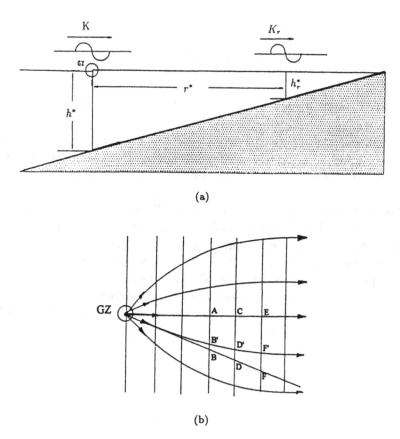

(b)

Fig. 7.2. (a) Initially axisymmetric water waves due to an underwater explosion at h^* are incident on a beach with uniform slope S. (b) Schematic representation of the ray paths bending towards the shore line due to refraction.

which for small angles of incidences is

$$K_R = \left[\frac{r \, \sin(\alpha_0)}{\int_0^r \sin(\alpha_r) ds} \right]^{1/2} .$$

(7.60)

Taking $\sin(\alpha_0)$ to be the denominator, $\frac{\sin(\alpha_r)}{\sin(\alpha_0)}$ can be replaced by $(\frac{k}{k_r})$. The refraction coefficient at any r can be computed as

$$K_R = \left[\frac{r}{\int_0^r \frac{k}{k_r} ds} \right]^{1/2} . \tag{7.61}$$

For shallow water, when $k = \sigma/(gd)^{1/2}$, and at the shoreline, Equation (7.61) is reduced simply to $K_R = \sqrt{\frac{3}{2}}$.

It is interesting to note that even though the wave ray is perpendicular to bottom contour, the refraction coefficient is not equal to one but tends towards a larger value (7.61) for a wave originating at a point source on a plane sloped bottom.

Replacing dk by $d\sigma/V$ as done previously, Equation (7.54) is evaluated using FFT. K_S and K_R are computed for each value of σ. Every σ uniquely corresponds to a value of k computed iteratively. Figure 7.3 shows a comparison of a theoretical wave record on a 1/20 slope with wave record propagated on a plane bottom. The effects of shoaling and refraction are clearly seen. An important point is that maximum effect of shoaling/refraction is felt at the leading wave, where the solution obtained using the Stationary Phase method would have been of little use.

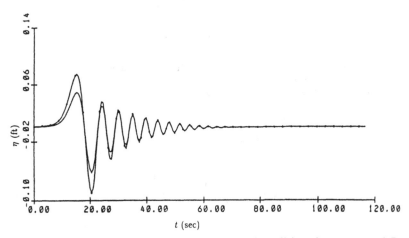

Fig. 7.3. Comparison of wave records for a "dome" initial condition of $\eta_{max} = 1$ and $R_c = 3$. The bottom slope $S = 1/20$. The dotted curve is the solution on slope at $r = 15$ from GZ, while the smooth curve is the solution on a flat bottom.

7. Experimental Verification of Propagation on 3D Bathymetries

A three-dimensional scale model of Sinclair Inlet was built at the Coastal Engineering Research Center, Vicksburg, Mississippi [Bottin *et al.*, 1988(a) and (b)]. The model was geometrically undistorted at a scale of 1/250 (time scale: 1:15.81). The model was equipped with a programmable snake paddle with 15 individual panels able to generate a EGWW-like transient wave. The programming of the paddles is discussed in the next chapter.

The waves were recorded in the scale model at many locations on the slope in front of the paddle, at the top of the slope where the natural bathymetry is reproduced to scale, inside the inlet and along the quays and wharfs near Bremerton (see Figure 7.4).

A hypothetical GZ was selected at a distance of 9.2 feet behind the wave generator along a perpendicular crossing the center of the wave paddles. This corresponds to a minimum prototype distance of 5300 feet between GZ and the beginning of the reproduced bathymetry.

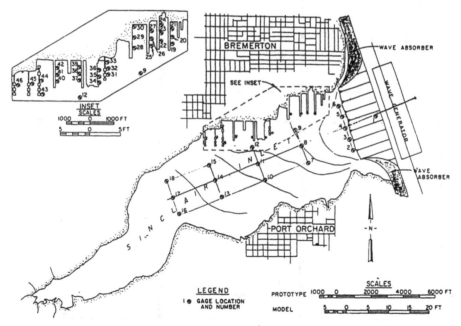

Fig. 7.4. Scale model of Sinclair Inlet.

The movement of the paddles was determined in pairs with respect to the medium axis, allowing the reproduction of a transient wave with radial symmetry. Theoretical curves have been determined by the application of the methods developed in this chapter with emphasis on the leading waves. It accounts for linear shoaling, refraction, and radial dispersion. These theoretical curves have been compared to experimental curves, obtained from the corresponding experiments.

In general the experimental curves provide a good verification of the theoretical calculation for the leading waves, as shown in Figures 7.5 and 7.6.

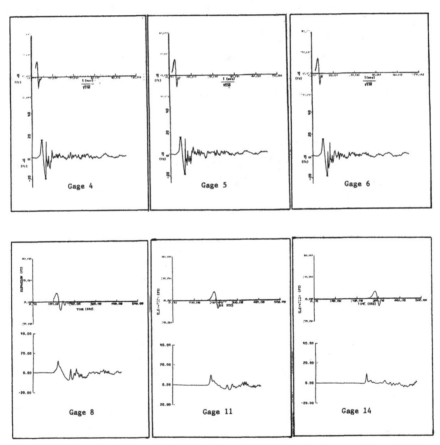

Fig. 7.5. Comparison of recorded wave with predicted wave for the case $W = 100$ KT.

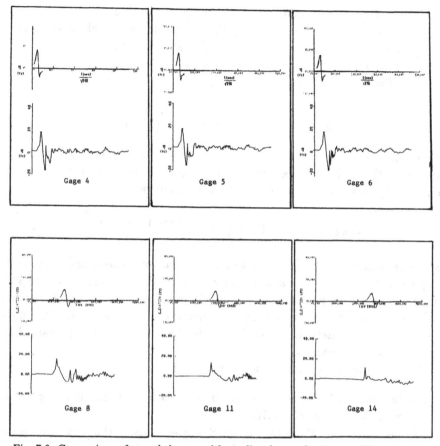

Fig. 7.6. Comparison of recorded wave with predicted wave for the case $W = 250$ KT.

However, by simply looking at the experimental curves, it appears that they are affected by a number of nonlinear effects which the linear theory cannot describe adequately. The experimental curves are steeper and higher than their theoretical linear equivalents. This is particularly true for the wave gages located in very shallow water.

8. Nonlinear Shoaling

The introduction of nonlinear effects compounds the difficulties by orders of magnitude. Nevertheless, in order to obtain more realistic results, the introdution of some correction factors due to nonlinear effects is possible.

In shallow water, before breaking, the waves suddenly peak in amplitude. This phenomenon is well explained by nonlinear shoaling and has been experimentally verified. Therefore, for better accuracy, it is possible to replace K_S by a nonlinear shoaling coefficient K_{SNL} (Iwagaki, 1968; Iwagaki and Sagai, 1971).

$$K_{SNL} = K_S + 0.0015 \left(\frac{d}{L_0}\right)^{-2.8} \left(\frac{H_0}{L_0}\right)^{1.2}, \tag{7.62}$$

where L_0 is the linear deep water wave length, d is the local water depth, and H_0 is the equivalent deep water wave height obtained by linear shoaling. $H_0 = H/K_S$ which can be expressed in dimensionless terms as

$$K_{SNL} = K_S + 0.0015 \left(\frac{\sigma^2}{2\pi}\right)^{-1.6} \left(\frac{H_s}{K_S}\right)^{1.2} \left(\frac{1}{d_s}\right)^{2.8}, \tag{7.63}$$

where H_s and d_s are the dimensionless local linear wave height and depth with respect to the water depth d at the origin. Since the shoaling coefficient is a function of the local wave height, the calculations have to be done step by step along wave rays. Since the wave rays are functions of frequency, this can only be done approximately, where nonlinear shoaling is the most important, i.e., prior to wave breaking.

The following empirical method is proposed:

(1) Determine the linear wave height till they reach the breaking index curve, such as defined by $H_b = 0.78d$.
(2) Come back along the wave ray which corresponds to the associated zero-crossing frequency, to a value of $d/L_0 \cong 0.2$ to 0.05 where the nonlinear effect is small.
(3) Then calculate the nonlinear value of K_S between the above d/L_0 and the breaking index curve by successive approximation.

This method has been introduced in the calculation of Figures 1.23 to 1.31.

This method, even though empirical, allows a much more accurate definition of the wave height before breaking, and of the extent of the surf zone. What is noticeable is that all the experimental leading waves exhibit a higher amplitude than their theoretical equivalent despite the fact that energy dissipation is neglected. This fact is in part attributed to neglecting nonlinear shoaling.

Overall, it can be concluded that the theoretical results on the propagation of the leading waves of a transient wave is relatively well verified when compared with experimental results. Both the theory and the experiments suffer from inaccuracies so that a thorough comparison is difficult. However, both methods appear adequate to predict the wave environment generated by impulsive source with an accuracy which is compatible with engineering purposes. For consistency, a method for nonlinear wave refraction should be used. This is beyond the state of the art in the case of a transient wave.

Chapter 8

LABORATORY SIMULATION
OF EGWWs

1. The Need for Scale Model Simulation

The propagation of transient waves over complex three-dimensional bathymetries presents a challenging mathematical and numerical problem as shown in the previous chapter. Nonlinear convective and dissipative effects are difficult to resolve or even to formulate mathematically and these effects are often too important to be neglected. Such is the case of the wave dissipation processes taking place along the banks of an estuary or along the wharfs of naval bases.

However, all these effects can be reproduced in similitude in physical models. The nonlinear inertial effects obey the Froude similitude. The short turbulent dissipative processes, such as those due to wave breaking, obey the generalized Froude similitude (Le Méhauté, 1991). The scale effects on the wave propagation are limited to the viscous boundary layer and capillary forces. These can be minimized by selecting an appropriate scale. For example, the scale model wave period should exceed 0.3 seconds, and the minimum water depth should exceed 2 inches. Under these conditions, the physical scale model becomes a better simulator than its mathematical counterpart.

Therefore, the idea of investigating the propagation and effects of EGWWs inside a naval base of complex configuration and bathymetry experimentally in preference to or in parallel with mathematical modeling makes sense. Theoretically, the problem would then consist of determining the size of the HE explosion for the laboratory scale model, which would reproduce the prototype EGWWs in similitude.

In order to reproduce EGWWs at a small scale, three methods could be used. The first one is evidently to scale down the explosion. Even though this method is not practical, that possibility is examined. The second method consists of using a programmable snake paddle. The third consists of simulating EGWWs by dropping a circular plunger or a circular plate into the water. This method is particulary convenient to simulate the effect of EGWWs at a short distance from GZ.

2. Scale Model Simulation of EGWWs by Small Charges

Can a small yield of TNT, exploded in a scale model of a body of water, simulate small EGWWs which are in similitude with large EGWWs generated by a large yield in the prototype?

First, since EGWWs are gravity waves, the governing similitude is Froude. Therefore, given a geometrical ratio λ between dimensions of the scale model and the prototype, the ratio for depths d free-surface elevation, η, horizontal distances and in particular r, must equal λ and the ratio for time scale is $\lambda^{1/2}$:

In order to achieve this, it suffices that the ratio of dimensions of the initial disturbances also equal λ. This leads to a very simple law in the case of shallow water, $(d/W^{1/3} < 1)$. Indeed, in this case, as seen in Chapter 3, EGWWs are practically independent of the depth of burst z. Furthermore, in Chapter 5, it was seen that the radius of the physical crater is proportional to $W^{1/4}$ (5.6).

For similitude, since $(d_m/d_p) = \lambda$, one must have $(R_{cm}/R_{cp}) = \lambda$. The subscripts m and p refer to model and prototype respectively. Therefore, $(W_m/W_p)^{1/4} = \lambda$. The yield scales to the one-fourth power of the length scale. For example, a 1-MT explosion in 100 feet of water generates EGWWs, which can be simulated by 20 pounds of TNT in one foot of water (scale: 1/100). The turbulent energy dissipation processes taking place near GZ are in similitude by virtue of the generalized Froude similitude, the same way that a small hydraulic jump (or tidal bore) could be in similitude with a large hydraulic jump. The fine structure of the flow, size of air bubbles, etc., are not reproduced but the total amount of energy dissipated is governed by application of the momentum theorem, which is in similitude.

The scale model wave records $\eta_m(t)$ at a distance r_m provides us with the prototype wave record $\eta_p(t)$ at a distance $r_p = \frac{r_m}{\lambda}$, by dividing η_m by λ ($\eta_p = \frac{\eta_m}{\lambda}$). The time scale t_p is divided by $\sqrt{\lambda}$ ($t_p = \frac{t_m}{\lambda^{1/2}}$).

In the deep water case ($\frac{d}{W^{1/3}} > 16$), the problems of simulation is much more complex, as the depth of burst (and atmospheric pressure) intervenes. The linear equivalent radius R is given by a complex function which is not amenable to a consistent similitude relationship. This is even more evident when considering the deep water case. Recall

$$\frac{\eta r^\alpha}{W^\beta} = C\left(\frac{d}{W^{1/3}}\right) \tag{8.1}$$

When $d/W^{1/3} > 14$, $\alpha \to 1$, and $\beta \to 0.58$.

Since ηr is proportional to $\eta_{0\,max} R$, if η (and η_0) is proportional to $W^{1/4}$, it follows that R has to be proportional to $W^{1/3}$ ($1/4 + 1/3 = 0.58$). The equivalent initial disturbance defined by the two parameters $\eta_{0\,max}$ and R, do not follow the same power of W. For a given relative depth of burst, a small yield causes a relatively deep narrow equivalent linear crater, whereas a large yield causes a relatively large shallow linear crater. In a word, the small crater is a vertically "distorted" model of the large crater, but distortion is not allowed by the dispersion relationship. So that even if the amplitude is in similitude (by keeping $\frac{\eta_m}{\eta_p} = (\frac{W_m}{W_p})^{1/4}$), the wave length, wave period, and complete time history of the wave train is not. If $\frac{R_m}{R_p} = (\frac{W_m}{W_p})^{1/3}$, the locations of wave crests, troughs and zero crossing on the time axis should be in similitude, but the amplitudes are not. One may wonder whether one cannot simulate EGWWs by "distorting" the depth of burst. It is possible but the state of the art of establishing the relationship between wave and explosion parameters, as given in Chapter 3, does not allow us to determine it. If it were, the calibration explained in Chapter 3 would have been much simpler.

As a conclusion, EGWWs can be simulated quantitatively in a scale model only in the shallow water case ($d/W^{1/3} < 1$). We can then expect that the hydrodynamic energy dissipation processes due to the bore are also in similitude. Actually, the problem of simulation of EGWWs in the laboratory by explosion is rather academic, since the use of an HE explosive, even when small, is not practical. The generation of EGWWs by snake paddle is now examined next. Initially, the simpler case of monochromatic waves at an angle is presented.

3. Generation of Monochromatic Waves at an Angle by Snake Paddle

EGWWs can also be simulated by a programmable continuous snake piston-type paddle located between GZ and the area where the wave propagates. Such a system has been described by Outlaw (1984). The paddle described by Outlaw is made of 60 elements, — individually controlled at the joints. The elements are $\bar{p} = 0.46$ meters wide and 0.76 meters high. They are designed to operate in $d = 0.60$ meters of water (2 feet). (See Figures 8.1 and 8.2.)

In the one-dimensional case, the maximum wave height which can be generated is a function of the maximum horizontal stroke $2e_m$ of the wave paddle and the period. For nonbreaking monochromatic waves, the wave height is given by Biesel and Suquet (1951) as

$$H = e_m P(kd) = e_m \frac{4 \sinh^2 kd}{\sinh kd \cosh kd + kd} \tag{8.2}$$

Fig. 8.1. Radial wave pattern using four generator modules (photo by Waterways Experiment Station).

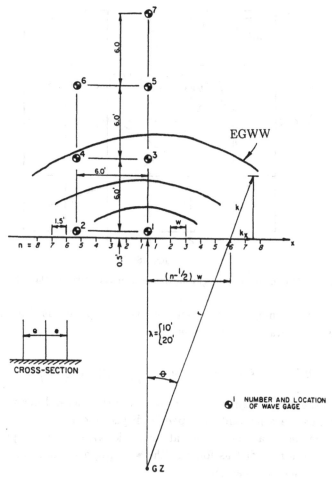

Fig. 8.2. Location of impulsion generated wave source, individual wave paddles, and wave gages.

where k is the wave number. The wave paddle also generates an infinite number of harmonic perturbations which decrease exponentially in distance from the paddle.

For small periods, the wave height is limited by the wave breaking conditions at the paddle and the maximum wave paddle velocity is mechanically achievable. The corresponding limits are presented in Figure 8.3. If the wave

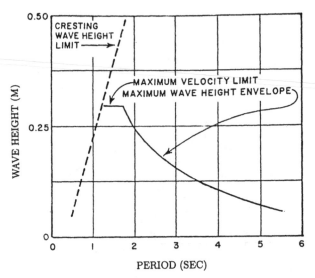

Fig. 8.3. Maximum wave height for monochromatic wave (Outlaw, 1984).

movement is harmonic, the movement of the wave paddle is sinusoidal, then the free-surface elevation $\eta(t)$ is given by:

$$\eta(t) = \frac{H}{2} \sin \sigma t = e_m \cos \sigma t \, \frac{2 \sinh^2 k^*}{\sinh k^* \cosh k^* + k^*} \, , \qquad (8.3)$$

where H is the wave height and $k^* = kd$. Note the phase difference between the free-surface elevation and wave paddle displacement.

When the waves are generated at an angle with the wave paddle axis, Equation (8.2) is modified as follows (the $*$ is dropped from here on for the dimensionless wave number k):

$$H = e_m \, \frac{k}{(k^2 - k_x^2)^{1/2}} \, \frac{4 \sinh^2 k}{\sinh k \cosh k + k} \, , \qquad (8.4)$$

where k_x is the wave number along the wave paddle, i.e.,

$$k_x = k \sin \theta \, ,$$

where

$$\cos \theta = \frac{\ell}{[\ell^2 + (n\bar{p})^2]^{1/2}} \, , \qquad (8.5)$$

where ℓ is the shortest distance from GZ to the wave paddle, and $n\bar{p}$ is the distance between the perpendicular to the wave paddle passing through the GZ and the paddle element joint number n (Figure 8.2). The width of a wave paddle element is $\bar{p}$.

The maximum possible wave angle without the generation of spurious waves is given in Figure 8.4.

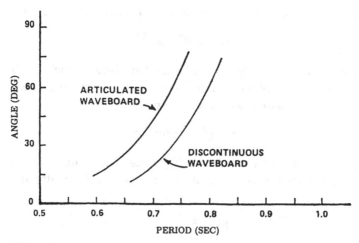

Fig. 8.4. Wave generators directional angle limits without generation of spurious waves (Outlaw, 1984).

Accordingly, combining Equations (8.3) and (8.5) gives

$$\eta(t) = \frac{H}{2} \sin \sigma t = e_m \cos \sigma t \, \frac{[\ell^2 + (n\bar{p})^2]^{1/2}}{\ell} \, \frac{2 \sinh^2 k}{\sinh k \cosh k + k} \, . \tag{8.6}$$

The generation of transient waves is presented next.

4. The Generation of 3D Transient Waves by Snake Paddle

The theory of the generation of three-dimensional transient waves by piston-type continuous snake paddle is now established (Le Méhauté *et al.*, 1989). Since the movement is linear and composed of an infinite number of harmonic components, the transfer function which relates the free-surface elevation (n, t) to the stroke $e(n, t)$ given by (8.6) is applicable to each component of the spectrum. Assume for example that the free-surface elevation at each wave paddle is given by one of the solutions for (r, t), then

$$\eta(r, t) = d\eta(r, t) = dA \int_0^\infty kdk \, H(k) J_0(kr) f\left(t + \frac{\pi}{2}\right) f(\theta) p(k) \qquad (8.7)$$

where $H(k)$ is the Hankel Transform of the initial disturbance. For example, in the case of a vertical displacement in the form of a mound (see Chapter 2):

$$H(k) = R^2 \exp^{-kR} . \qquad (8.8)$$

A and R characterize the size of the initial disturbance which can be related to the yield and depth of burst. Parameter r is the distance between GZ and one edge of a wave paddle element given by

$$r = [\ell^2 + (n\bar{p})^2]^{1/2} , \qquad (8.9)$$

and $f(t) = \sigma \sin \sigma t$ for an impulse or $f(t) = \cos \sigma t$ for a dome or crater; then the movement of each wave paddle is obtained straightforwardly as:

$$e(n, t) = \frac{A\ell d}{[\ell^2 + (n\bar{p})^2]^{1/2}} \int_0^\infty kdk \, \frac{\sinh k \cosh k + k}{2 \sinh^2 k}$$

$$\times J_0\left[\frac{k}{d}(\ell^2 + (n\bar{p})^2)^{1/2}\right] \bar{H}(k) \left| \begin{array}{c} \sigma \cos \sigma t \\ \sin \sigma t \end{array} \right| . \qquad (8.10)$$

This formulation is exact within the approximation of the theory, which is linear and neglects the terms which decay exponentially in distance from the wave paddle (Biesel and Suquet, 1951). If the wave paddle is located sufficiently far away from GZ, the integral form can easily be replaced by using the stationary phase approximation as in Chapter 2. Otherwise dk can be replaced by $\frac{d\sigma}{V}$, and the integral is determined by FFT.

5. Experimental Verification of the Snake Wave Paddle Generation

An experimental program of research has been carried out at the Coastal Engineering Research Center, Waterways Experiment Station, to verify the previous theory. The experiments were carried out in a 2-feet deep basin with a section of snake paddle composed of 16 elements (Figures 8.1 and 8.2).

The movement of each wave paddle $e(n, t)$ is calculated from Equation (8.10) assuming radial symmetry from a GZ located on a center line perpendicular to the wave paddle at a distance ℓ of 10 feet and 20 feet, successively.

Figure 8.5 provides an example of the movement $e(t)$ of wave paddle corresponding to formula (8.10). The wave-type histories were recorded at seven locations in the basin where the wave was initially determined theoretically from Equations (8.7) and (8.8).

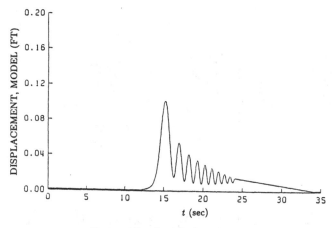

Wave paddle displacement for λ = 10 feet

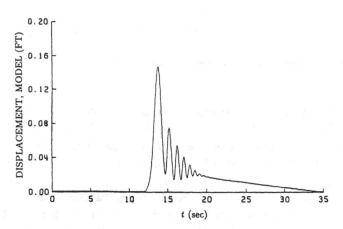

Wave paddle displacement for λ = 20 feet

Fig. 8.5. Wave paddle movement.

The experimental results provide a good verification of the theory for all the gages (Figure 8.6). However, they indicate that the amplitude of the recorded wave is slightly smaller than predicted, a fact which can easily be explained by the energy losses due to friction and the leaks between the paddles and the bottom of the tank. The discrepancies are larger for the leading waves, which can also be attributed to the mechanical limitation (maximum possible acceleration) of the wave paddle.

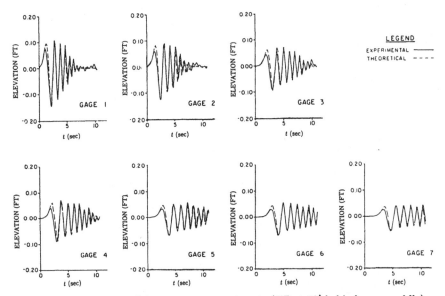

Fig. 8.6. Comparison of theory versus experiments (GZ at 10′ behind wave paddle).

By multiplying the experimental results by a factor of 1.2 (corresponding to a paddle efficiency of 83%), the match between the theoretical and experimental results becomes excellent, as seen in Figure 8.6, which corresponds to tests where the hypothetical GZ is at a distance of 10 feet.

The small irregularities which appear for the gages at a distance from the paddle (i.e, gage 7 in Figure 8.2) are due to the vagaries of refractive effects. This results from the fact that the basin, theoretically horizontal, actually introduces some very small irregularities in the water depth, which are sufficient to influence the wave amplitude at a distance.

Identical tests have been carried out by varying the wave amplitude, or the factor A in Equation (8.10) by a factor of 2. The corresponding experimental results are presented in Figure 8.7. To match theoretical and experimental results requires a multiplying factor of 1.2, which indicates that the efficiency of the wave paddle is little affected by nonlinear effects.

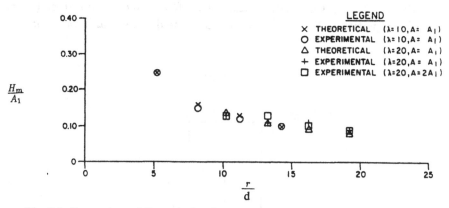

Fig. 8.7. Comparison of theoretical and experimental maximum wave height data.

Equation (8.10) is valid when the water depth at GZ d, at the wave paddle d_p, and in the model d_m are all the same. This has been the case in the experiment previously described.

For practical applications to specific locale, the bathymetry is in general three-dimensional, and the depth at GZ can be different from the depth at the limit of the model where the three-dimensional bathymetry is reproduced. Also the wave paddle may be located in deeper water than that which exists in the prototype. Then a number of transformations are needed.

In order to solve this problem, the theory for the propagation of transient waves over a three-dimensional bathymetry developed in the previous section is applied.

6. Generation of a Transient Wave with Axisymmetry in a 1D Wave Flume

The effect of EGWWs on structures with axisymmetry can also be investigated in a narrow one-dimensional wave flume, equipped with a programmable

piston-type wave paddle, provided only a narrow segment needs to be reproduced. This procedure generally allows experimentation at a larger scale than
in a two-dimensional basin. Referring to Figure 8.8, the problem now consists of determining the time history of the paddle movement at $x = 0$ in the
one-dimensional wave tank which will simulate the same wave time history at
$x = x_1$ and the axisymmetry wave at a distance $r = r_1$ from a hypothetical
source center at GZ. Consider, for example, a free-surface time history (r_1, t)
given by Equation (8.7) where $r = r_1$. In a one-dimensional wave tank, the

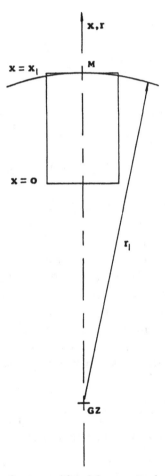

Fig. 8.8. Generation of a wave with axial symmetry in a narrow wave flame.

time history at a distance $x = x_1$ is given by replacing the Bessel function $J_0(kr)$ by $\cos(kx_1)$ as follows:

$$\eta(x_1, t) = \int_0^\infty dk \cos kx_1 \cos \sigma t F(k). \tag{8.11}$$

$F(k)$ defines the original disturbance at $x = 0$, which is unknown. Replacing dk by $d\sigma/V$, where V is the group velocity, Equation (8.11) is recognized as a Fourier Transform of:

$$G = \frac{F(k) \cos kx_1}{V}, \tag{8.12}$$

such as:

$$\eta(x_1, t) = \int_0^\infty G \cos \sigma t \, d\sigma. \tag{8.13}$$

Replacing $\eta(x_1, t)$ by $\eta(r_1, t)$ and taking the reverse transform yields:

$$G = \frac{F(k) \cos kx_1}{V} = \frac{2}{\pi} \int_0^\infty \eta(r_1, t) \cos \sigma t \, dt \tag{8.14}$$

which can be integrated numerically. Then $F(k)$ is determined:

$$F(k) = \frac{VG}{\cos kx_1}. \tag{8.15}$$

Inserting Equation (8.15) into (8.11) the wave elevation at any location x in the wave flume is:

$$\eta(x, t) = \int_0^\infty dk \cos kx \cos \sigma t \frac{VG}{\cos kx_1}. \tag{8.16}$$

Considering the location of the wave paddle $x = 0$, the wave paddle displacement $e(t)$ was able to generate $\eta(x_1, t)$, or equivalently $\eta(r_1, t)$ can be found by using the transform operator given in Equation (8.2):

Then,

$$e(t) = \int_0^\infty dk \sin \sigma t \frac{VG}{\cos kx_1} \frac{\sinh k \cosh k + k}{2k \sinh^2 k}, \tag{8.17}$$

which can also be integrated numerically. The response of a moored submarine to EGWWs has been investigated in a large two-dimensional wave tank by using this technique.

7. The Use of a Plunger

For the near field, i.e., to investigate the effects of EGWWs on structures and ships near GZ, it is not possible to use the snake paddle. Instead, dropping a plunger plate at GZ can be used. Simulation of EGWWs has been done by making use of a variety of plungers which were dropped and raised rapidly (Jordaan, 1964). The plungers included a cylinder, a flat disk, a paraboloid, and even an air jet striking the water has been used. In a series of tests, a 14-foot diameter, 3-foot deep paraboloid plunger was used. No attempt was made to relate its physical characteristics to yield. It did provide a reasonably good verification of the early theory of Kranzer and Keller (1959) on impulsive wave theory.

The dropping of a circular plate for simulating EGWWs has been used in foreign laboratories and in the US. For example, the simulation of EGWWs at a scale of 1/150 in the Naval Base of Mers-el-Kebir was done by this method in 1952 in the hydraulic laboratory of Sogreah, Grenoble. Earlier, Johnson and Bermel (1949) investigated the wave run-up due to the Bikini explosion on a scale model of the island in the Berkeley Hydraulic Laboratory, also by dropping a cylindrical plate.

In both cases, little research was done on the sizing of the physical characteristics of the plate as a function of yield and scale. Also little was known at that time about the magnitude of EGWWs which have to be simulated. Nevertheless, it was concluded that a plate with a vertical edge, causing vertical splashing at its perimeter, would be a better simulator than a plate with a tilted or a curved edge in shallow water.

In view of what we have learned in Chapters 4 and 5 and in this chapter, the sizing of the crater radius as a function of the yield to be simulated as a function of scale is straightforward in the shallow water case. Indeed, we know that the radius of the water crater caused by explosion in shallow water at any scale is: $R_c = 4.4\,W^{0.25}$. Therefore, any mechanical means allowing a rapid creation of a physical crater should have the same wave generation potential as its explosion counterpart. This is most likely to be achieved by dropping a plate of the same radius $R_p = R_{cm} = R_{cp} = W_p^{1/4}$. However, since the plate remains on the bottom of the scale model after it has sunk, it must be very thin to allow for the free collapse of the water crater surrounding it. Indeed, referring to Chapters 4 and 5, a thick plate would cause a longer time to elapse between the leading wave created by the original splash and the following wave due to crater collapse than in the case of EGWWs. This is due to the fact that the water depth becomes smaller at the location of the plate and wave velocities decrease as depth decreases. The dynamics of a falling plate in the water is now analyzed.

8. Slamming of a Falling Plate in Shallow Water

When a flat circular plate (or a vertical cylinder) is dropped on the free surface of a quiescent body of shallow water, the following physical phenomena take place. It is assumed that the plate hits the surface flat with perfect symmetry. Initially, a large slamming force occurs as evidenced by the splashing at the plate's edge. A compression pressure wave is transmitted towards the seafloor at the speed of sound in water and reflected back and forth. Assuming that the water depth is small compared to the plate radius, the entire mass of water under the plate is rapidly subjected to an initial increase of pressure uniformly. Assuming that the plate is perfectly rigid, the initial slamming pressure is $p_0 = \rho w_p c$ where ρ is the water density, w_p is the plate's impact velocity ($w_p \cong (2gz)^{1/2}$), g is the gravitational constant, z is the height from where the plate is dropped and c is the sound speed in water. The force due to this pressure p_c causes a considerable deceleration of the plate at impact. For example, assuming $\rho = 2$ slug/ft.^{3}, $c = 4,500$ ft./sec., and $z = 6$ feet, then $w = 19.65$ ft./sec. and the total force F on a 6-foot diameter plate is approximately $5 \cdot 10^6$ pounds. This causes deceleration of a 1-inch thick, 6-foot diameter rigid steel plate (weighing 1062 pounds) on the order of

$$\frac{dw}{dt} = -\frac{F}{M_p} - g \cong -5 \cdot 10^3 \, g, \qquad (8.18)$$

where M_p is the mass of the plate. (The elastic property of the plate intervenes to reduce the value of the deceleration so that the plate does not slow down as much.)

This magnitude of deceleration is difficult to record experimentally. It will be shown that the response of the accelerometers used in the experiments were unable to provide a quantitative verification of Equation (8.18) due to slamming.

The large increase of pressure $p_0 = \rho w_p c$ which takes place uniformly between the plate and the concrete floor is now released at the edge of the plate, where the pressure is initially equal to atmospheric pressure (plus hydrostatic). A rarefaction wave travels in the water from the edge to the center of the plate at a velocity a, (Figure 8.9) which is dependent on both the fluid compressibility and the elastic properties of the plate and the concrete floor. This is analogous to the theory of a water hammer in a pipe. The pressure wave velocity in a steel pipe is around 3000 ft./sec. and is much smaller in a concrete pipe.

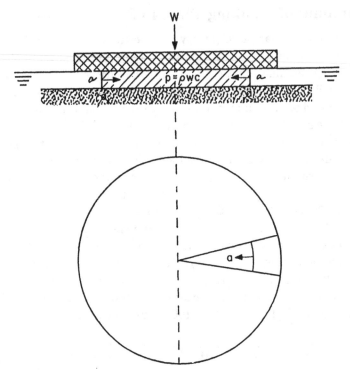

Fig. 8.9. Initial water pressure under a plate falling on shallow water.

The velocity a is even smaller when one of the boundaries is a thin concrete slab built on a sandy foundation as is the case in laboratory scale experiments. Assume for example that it is 600 ft./sec. Then it takes a time period, $\tau = R_p/a = 3/600 = 0.005$ seconds, where R_p is the radius of the plate, for the rarefaction wave to reach the center of the plate. The exit water velocity at the edge of the plate, where the pressure is initially p_0, is

$$U_R = \frac{p_0}{\rho a} = w_p \frac{c}{a} \cong 7.5 w_p. \tag{8.19}$$

If the value of w given in the previous section is used, then $U_R \cong 150$ ft./sec. This horizontal outward movement is deflected vertically by the water inertia at a short distance from the plate edge causing the conspicuous vertical splashing.

As the rarefaction wave travels towards the center of the plate, water is expelled and the pressure decreases. Since the cross section decreases with the radius, there is a "whip effect" like in penstock and pipeline of converging cross sections (Figure 8.9), i.e., the pressure keeps decreasing. As the pressure decreases it eventually reaches vapor pressure p_v, which causes the water under the plate to cavitate. The outward water velocity at the front of the pressure wave is then

$$U_r = \frac{p - (-p_v)}{\rho a} = w_p \frac{c}{a} + \frac{p_v}{\rho a} = 150 + \frac{30}{1200} \cong 150 \,. \tag{8.20}$$

For continuity $U_r = U_R(R^2/r^2)$ where r is the horizontal radial distance from the center of the plate. Because the plate is subjected to atmospheric pressure p_a on the top and the cavitation pressure on the other side, its acceleration becomes positive again and larger than g:

$$\frac{dw}{dt} \cong \frac{\pi R_p^2}{M_p}(p_a + p_v) + g \cong \frac{2p_a}{\rho_s H_p} + g \,, \tag{8.21}$$

where ρ_s is the plate's density and H_p is the plate thickness. In the case of a one-inch thick plate, the acceleration is then of the order of $3g$.

After a time $t = t_c = \frac{2R}{a} \cong 0.01$ sec., the negative wave causing cavitation is replaced by a positive wave which causes a deceleration of the plate to a value smaller than g. The analysis can easily be pursued further, causing successively acceleration and deceleration. However, because of energy dissipation at the edge of the plate, the water hammer effect is rapidly damped. The downward movement of the plate is then governed by the hydrodynamic forces of an incompressible fluid. Figure 8.10 presents a schematic time history of pressure and water velocity under a plate falling in shallow water.

Once the compressible phase is over, the plate decelerates rapidly. The value of the velocity can then be approximated by the balance of forces as follows. Let w_0 be the velocity of the plate at the end of the compressible phase. It is assumed that w_0 is of the same order of magnitude as the velocity when the plate reaches the free surface. (The deceleration due to slamming is very large, but it is exerted over an extremely short period of time.) Therefore, if one assumes the fluid to be incompressible, then when the plate hits the free surface, its momentum is suddenly reduced by $\Delta M = M(w - w_0)$. This change of momentum is equal to the impulse resulting from the pressure forces $p(r, \theta)$ (where θ is the angle) exerted vertically by the water on the plate, namely,

$$\Delta M = \int_0^{2\pi} \int_0^{R_p} \int_0^{t_0} p(r, \theta) r \, dr \, d\theta \, dt_0 \,, \tag{8.22}$$

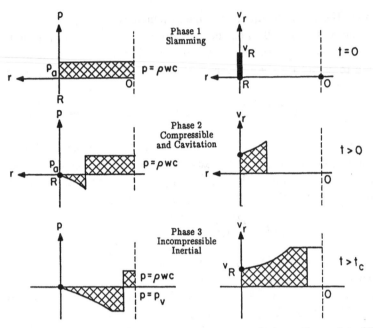

Fig. 8.10. Schematic time history of pressure and water velocity under a plate falling on shallow water.

where t_0 is the duration of the initial phase $(t_0 = O(\varepsilon))$. Here, dt can be considered as a Dirac delta function δt_0.

The same vertical pressure forces are also exerted by the cylinder on the water, which is expelled inducing a horizontal movement analogous to a vertical jet hitting a horizontal plate. Since the hydrostatic forces are cancelled, application of the momentum theorem to the volume of water under the cylinder gives:

$$\Delta M = \int_0^d \int_0^{2\pi} \int_0^{R_p} \rho u_r\, r dr\, d\theta\, dz \qquad (8.23)$$

where u_r is the horizontal radial velocity and d is the water depth. Let $\bar{h}$ be the distance between the plate and the bottom $(\bar{h}|_{t_0} = d)$. Then by the equation of continuity

$$\pi r^2\, d\bar{h} = -2\pi\, r d\, u_r\, dt\,, \qquad (8.24)$$

and $d\bar{h}/dt = -w_0$, so that

$$u_r = \frac{r w_0}{2D}\,. \qquad (8.25)$$

Inserting this into the integral (8.23) gives:

$$M(w - w_0) = \frac{\rho \pi w_0 R_p^3}{3},$$ (8.26)

or defining $\chi = \rho \pi R_p^3 / 3M = \rho R_p / 3 \rho_s H_p$, then

$$w_0 = \frac{w}{1 + \alpha},$$ (8.27)

and

$$\Delta M = M w \frac{\chi}{1 + \chi}.$$ (8.28)

Applying this formula then one finds: $\chi \cong 1$ and $w_0 \cong \frac{w}{2} = 8.8$ ft./sec. The plate velocity is approximately reduced by half during the initial impact.

9. Establishment of the Equation of Movement of a Sinking Plate

Once the compressible fluid phase is over and the plate speed velocity has been reduced, from a velocity w to a velocity w_0, the plate almost instantaneously penetrates the water and is subjected to time dependent pressure forces forcing the plate to decelerate. With h being the distance between the bottom and the lowest plate cross section, the cylinder's deceleration is

$$\frac{dw}{dt} = \frac{d^2 \bar{h}}{dt^2} = \bar{h}_{tt},$$ (8.29)

which acts in the same direction as gravity. Therefore, its inertia, together with its weight, causes a total force on the water

$$F = M(\bar{h}_{tt} + g) = \int_0^{2\pi} \int_0^{R_p} p(r) r \, dr \, d\theta,$$ (8.30)

where $p(r)$ is the sum of three components.

The first component is the hydrostatic pressure p_H given as

$$p_H = \rho g (d - \bar{h}).$$ (8.31)

Then the next pressure component p_w is due to the gravity wave components generated by the plate's downward movement as the water is squeezed between the plate and the bottom. (It is assumed that the large increase of pressure due to the initial impact recedes so rapidly than it could be characterized by the Dirac delta function as in the previous section.) In compliance with the linear long wave theory, pressure due to wave generation is hydrostatic, i.e.,

$$p_w = \rho g \eta = \rho g u_{wr} \left(\frac{d}{g} \right)^{1/2} , \qquad (8.32)$$

η being the free-surface elevation along the cylinder and u_{wr} being a fictive wave's horizontal velocity component. The velocity u_{wr} is related to the radial outwardly distance, taking place at the limit of the plate u_{wr} as follows:

$$\int_0^{d+\eta} u_{wR} dz \cong u_{wR} D = \int_0^{\bar{h}} u_R d\bar{h} = u_R \bar{h} , \qquad (8.33)$$

where u_R is the horizontal fluid velocity under and at the limit of the plate. Also from the equation of continuity

$$\pi R_p^2 d\bar{h} = -2\pi R_p \bar{h} u_R dt , \qquad (8.34)$$

thus

$$u_R = -\frac{R_p \bar{h}_t}{2h}$$

where

$$h_t = \frac{d\bar{h}}{dt} (h_t < 0) . \qquad (8.35)$$

Combining Equation (8.32) to (8.35) yields

$$p_w = -\frac{\rho R_p}{2} \left(\frac{g}{D} \right)^{1/2} \bar{h}_t . \qquad (8.36)$$

Finally, the increase of pressure under the plate also results from the inertia of the water which is squeezed out radially between its lower section and the bottom. Because this increase in pressure $p_I(r)$ acts along a vertical plate of area $2\pi r h$, the total force on the plate of radius r to the inertia of the water between r and R is:

$$2\pi r \bar{h} p_I(r) = \rho \frac{d}{dt} \int_0^{2\pi} \int_r^R \int_{h(t)}^0 u_r d\bar{h} \, r dr \, d\theta . \qquad (8.37)$$

Note that the mass of the water involved is time dependent through $h(t)$ as is the velocity u_r. From the equation of continuity

$$\pi r^2 d\bar{h} = -2\pi r \bar{h} u_r dt, \tag{8.38}$$

and thus

$$u_r = -\frac{r\bar{h}_t}{2\bar{h}}. \tag{8.39}$$

Inserting Equation (8.39) into (8.37) and rearranging the equation (since r does not depend upon t) yields

$$p_I(r) = \frac{\rho}{r\bar{h}} \int_{R_r}^{r} r^2 dr \frac{d}{dt} \int_{h(t)}^{0} \left(-\frac{\bar{h}_t}{2\bar{h}}\right) dh, \tag{8.40}$$

$$= \frac{\rho}{rh} \frac{R_r^3 - r^3}{3} \frac{d}{dt} \int_{0}^{t} \left(\frac{\bar{h}_t^2}{2\bar{h}}\right) dt, \tag{8.41}$$

$$= \frac{\rho}{2r} \frac{R_r^3 - r^3}{3} \frac{\bar{h}_t^2}{\bar{h}^2} \tag{8.42}$$

Inserting Equations (8.31), (8.36), and (8.42) into (8.30) gives

$$M_p(h_{tt} + g) = 2\pi\rho \int_{0}^{R_p} \left[g(d - \bar{h}) - \frac{R_p}{2} \left(\frac{g}{d}\right)^{1/2} \bar{h}_t + \frac{1}{2r} \frac{R_p^3 - r^3}{3} \left(\frac{\bar{h}_t}{\bar{h}}\right)^2 \right] r dr. \tag{8.43}$$

Therefore, the plate movement obeys the following differential equation

$$A\left[\frac{\bar{h}_{tt}}{g} + 1\right] = 1 - \frac{\bar{h}}{D} - B\bar{h}_t + C\left[\frac{\bar{h}_t}{\bar{h}}\right]^2, \tag{8.44}$$

where

$$A = \frac{M_p}{\rho\pi R_p^2 d}, \tag{8.45}$$

$$B = \frac{R_p}{2d(gd)^{1/2}}, \tag{8.46}$$

$$C = \frac{R_p^2}{4gd}, \tag{8.47}$$

with the initial conditions

$$t = 0, \quad \bar{h} = d, \quad \bar{h}_t = w_0. \tag{8.48}$$

Equation (8.44) can also be written as

$$\bar{h}^2 \bar{h}_{tt} + (p\bar{h}^2 + q)\bar{h}_t + s\bar{h}^3 = D \,, \tag{8.49}$$

where p, q, s, and D are coefficients related to A, B, and C.

10. Analysis and Results of a Sinking Plate

Equation (8.44) or (8.49) cannot be integrated analytically in general cases. Its numerical integration has not been done at this time. However, a number of partial solutions which are indicative of the solution can be obtained by simplifying assumptions. Neglecting the inertial forces of the water under the plate $(C(\bar{h}_t/\bar{h})^2$ yields a linear differential equation which can be integrated straightforwardly. However, the inertial term remains of paramount importance in determining the plate movement near the bottom. Therefore, this term should remain in order to determine how much of the plate's momentum is left when it reaches the bottom.

If one keeps this term, regardless of the other simplifying assumptions which are made for the sake of mathematical convenience, the solution always implies that $\bar{h}_t \to 0$ when $h \to 0$ (and also $h_{tt} \to 0$), i.e., the cylinder reaches the bottom with zero velocity.

Indeed when $h \to 0$, one finds that

$$t \to -\left(\frac{C}{A-1}\right)^{1/2} \ell n\left(\frac{\bar{h}}{d}\right) \,, \tag{8.50}$$

which tends to infinity when $h \to 0$, and

$$\frac{\bar{h}_t}{\bar{h}} \to \left(\frac{A-1}{C}\right)^{1/2} \,. \tag{8.51}$$

The escape velocity, which is given by Equation (8.35) where $r = R_r$ tends towards

$$u_R \to \frac{R_p}{2}\left(\frac{A-1}{C}\right)^{1/2} = \left(\frac{M - \rho\pi R_{pd}^2}{\rho\pi}\right)^{1/2} \frac{2q^{1/2}}{R_p} \tag{8.52}$$

when $h \to 0$, i.e., the plate reaches the bottom independent of the initial plate velocity w_0. Therefore, u_R remains finite.

These apparently paradoxical results can be explained *ad absurbo*. Indeed, if the plate's velocity h_t and acceleration h_{tt} were finite when the plate reaches the bottom, i.e., when $h \to 0$, then the velocity u_R will tend to infinity, and the pressure forces resisting the plate downfall will be infinite, as seen from Equation (24), forcing the velocity h_t to be zero.

Nevertheless, the time for the plate to reach the bottom is not infinite. It remains in practice that the plate eventually ends up steady on the bottom. In reality, the end effect is confined to a very small distance from the bottom. The flow under the plate loses its perfect radial symmetry before reaching the bottom. A very small tilting forces the plate to reach the bottom at its edge before being put to rest.

This phenomena can be observed by dropping a horizontal sheet of paper onto a flat surface. The paper seems to slip sideways on a very thin layer of air before coming to rest. This indicates that, first, it takes a relatively long time for the thin layer of air caught between the paper and the surface to escape, and second, the sideways movement indicates that the air escapes preferably on one side rather than another.

The important conclusion from this analysis is that, for practical purposes, the plate reaches the bottom with practically zero momentum. Therefore, all its fall momentum is eventually transformed into water waves. However, an important part of the corresponding energy is dissipated by splashing.

For the problem under consideration it also appears that the water waves generated during this second phase, the sinking phase, have little chance of resembling explosion generated waves such as described in previous chapters. This is in contrast with the initial pulse due to slamming described previously. Therefore, the leading waves which are the least influenced by the sinking phase could be reproduced in similitude by dropping a plate, but the trailing waves, which result from the combination of the two modes of generation, the impulse and the flow movement under the plate, cannot.

As the plate sinks to the seafloor, the water surges onto its top towards the center. The motion is akin to a surge on a dry bed or the "dam-break" problem as seen in Chapter 4. It is highly nonlinear. This radial convergence gives rise to a spike which collapses, forms a bore and radiates a second wave as is described in Chapter 4.

The leading wave generated by the initial impulse is not influenced by the sinking of the plate and the return flow. The trailing waves are the result of the combined effects of dispersion from the initial impulse, the waves generated during the sinking of the plate, and the return flow. Its complexity defies analysis. However, one should expect that the time duration between the leading waves and the second wave crest to be of the order of $T \cong 2R/(gd')^{1/2}$, where d' is the difference between water depth d and the plate thickness H_p : $(d' = d - H_p)$.

11. Experimental Verification of the Drop of a Flat Circular Plate

A series of tests have been carried out at WES with a 6-foot diameter, 1-inch thick steel plate falling over water of various depths (Bottin, 1990; Bottin et al., 1990). An accelerometer was mounted on the plate near its center. Some typical results are presented in Figure 8.11. The experimental results were marred by considerable elastic vibration of the plate, which caused larger accelerations than expected with the noise exceeding the signal. Nevertheless, the period of these vibrations was well determined from the experimental data obtained before the plate reached the free surface. Therefore, the data can easily be filtered. The average values of the acceleration support the theoretical developments done in the previous section.

The peak deceleration at slamming was recorded, but did not reach the magnitude predicted by the theory. This is probably due to the very slow response of the sensor. The acceleration during the cavitation phase reaches nearly $2g$ as predicted. As the plate sinks towards the seafloor, the vibrations

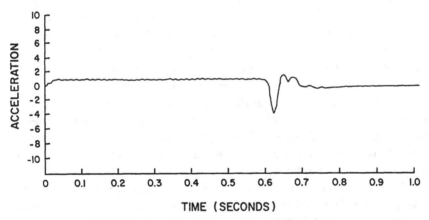

Fig. 8.11. Accelerometer recording of falling plate.

are damped. Then it appears very clearly that as the plate nears the seafloor, its deceleration becomes very small and tends nearly to zero when the plate reaches the seafloor as predicted by the theory.

12. Waves Generated by the Impulse of a Falling Plate

The problem of linear water waves generated by an impulse on the free surface is solved in Chapter 2. In the present case, the initial conditions are defined by an impulse on the free surface over a radius R equal to the plate radius and by a velocity $h_t(t)$ until the plate reaches the seafloor. The initial impulse I is

$$I \cong M(w - w_0) \, . \tag{8.53}$$

However, one has seen that the plate reaches the seafloor with zero momentum as the remaining momentum Mw_0 is nearly completely transmitted to the fluid during the sinking phase, so that practically $I \cong Mw$. Then, referring to Equation (2.23):

$$B(k) = \frac{Mwd^{-1}}{\rho \pi R^2} \int_0^R J_0(kr_0) r_0 dr_0 \, . \tag{8.54}$$

This integral is a Hankel Transform which can be integrated, analytically giving

$$B(k) = \frac{Mwd^{-1}}{\rho \pi R^2} \frac{R}{k} J_1(kR) \, . \tag{8.55}$$

Using the same methodology as the one presented in Chapter 2 yields:

$$\eta(r, t) = E \int_0^\infty J_0(kr) J_1(kR) \sigma \sin \sigma t \, dk \, , \tag{8.56}$$

where

$$E = \frac{Mw}{\rho \pi R_p} = \frac{\rho_s}{\rho} R_p H_p w \, . \tag{8.57}$$

Using the stationary phase approximation (Jeffreys and Jeffreys, 1956), which is valid for trailing waves at large distances r, gives after some arithmetic

$$\eta(r, t) = E \frac{\sigma D}{rk} J_1(kR) \left(\frac{kV}{-\dfrac{dV}{dk}} \right)^{1/2} \sin(\sigma t - kr) \, . \tag{8.58}$$

It is seen from Equations (8.56) and (8.58) that the wave heights are proportional to the fall velocity w. This should hold true for small values of w

within the confinement of the validity of the linear theory. In reality, non-linear effects are important and a maximum value for $\eta(r, t)|_{\max}$ is reached, corresponding to a maximum energy, after which the balance is dissipated by turbulent friction due to splash.

A number of experiments aimed at measuring water wave trains generated by the drop of a circular plate were done at the Coastal Engineering Research Center at WES. Typical wave records are shown in Figure 8.12, resulting from the drop of a one-inch thick steel plate 6 feet in diameter in 0.9 feet of water. As expected, the linear theory presented in the previous section is not verified due to the importance of nonlinear effects. However, this is of little practical importance. What really matters is whether the plate-generated wave and EGWWs look alike.

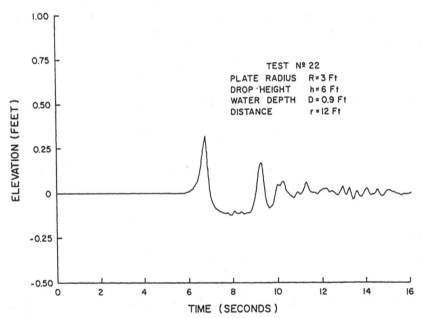

Fig. 8.12. Time history of the water waves produced by the drop of a six-foot diameter plate from a 6-feet height in 0.9 feet of water at a distance of 12 feet.

If our contention is right, i.e., if we assume that $R_c = R_p$, then according to Equation (5.6), this plate will be equivalent to an explosion of $(\frac{R_p}{4.4})^4 = (\frac{3}{4.4})^4 = 0.216$ pounds of TNT, also in 0.9 feet of water. (This 0.216 pounds of TNT is equivalent to a 10.8-KT (0.216 pounds $(100)^4$) explosion in 90 feet of water (scale = 1/100)).

A comparison between the two wave records (obtained by the plate and by the 0.216 pounds of TNT explosives) provides a direct verification of our assumption that $R_c = R_p$ or gives the appropriate proportionality coefficient. Unfortunately, this test was not done. However, we can compare the wave record obtained with 10 pounds of TNT exploding in 1 foot of water at a radial distance r of 10.2 feet (Figure 8.12) with a wave record obtained at a distance $r = 12$ feet with the drop of a 6-foot diameter plate in 0.9 feet of water (Figure 8.13). Despite the differences, which do not allow a direct quantitative verification, it appears that the main features of the wave train are identical: a large leading wave followed by a long trough, and an important second wave followed by secondary high frequency oscillations.

On the positive side, the sharp steepening of the wave front (and bore near ground zero), which is due to nonlinear convective effects, is reproduced experimentally whereas the linear theory gives too gentle a free-surface slope.

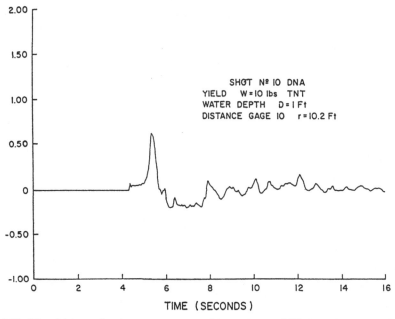

Fig. 8.13. Time history of water waves generated by a 10-pound TNT explosion in 1 foot of water at a distance of 10.2 feet from GZ.

As long as the main damage is caused by the leading wave, the drop of a plate for simulating the effect of EGWWs is a legitimate engineering tool and should provide realistic experimental results. However, if the damage is due to the trailing waves, the method is quantitatively invalid. We have seen that EGWWs can be reproduced relatively satisfactorily by a programmable snake paddle. But this method is not possible in the case of investigating the effects of an explosion near ground zero. Therefore, despite its deficiency, the simulation of EGWWs in shallow water by plate dropping appears to be the only possible method, short of a TNT explosion, to simulate a prototype nuclear explosion.

For a better correlation than the one presented in this chapter, more tests are necessary. A direct comparison of the wave record obtained by a small TNT explosion with the wave record obtained by dropping a plate of radius R_p equal to (or proportional to) R_c in the water of same depth d and at same distance r, would be the best way to proceed. For the rest (the extrapolation to large nuclear yield), one relies on the findings presented in Chapter 5.

Chapter 9

BIEM FORMULATION OF EXPLOSION BUBBLE DYNAMICS

1. Introduction

A numerical method to simulate the bubble expansion and crater formation resulting from an underwater explosion is presented in this chapter. It deals with the phase after the shock wave has been released and, therefore, all the energy in this phase is assumed to be going to wave formation without any other dissipation. The bubble expansion is described by a potential flow model in which the fluid is assumed incompressible and inviscid.

The correlation model of Wang *et al.* (1989 and 1991) provides a unified correlation for wave intensity predictions at any given distance from the explosion. On the other hand, this correlation is mainly based on small yield explosion tests. Extrapolation to large yield explosions is doubtful and uncertain. The logical approach to resolving this problem is to use a hydrocode. Several hydrocodes have been under development for a number of years showing very promising and interesting results. However, they still have not reached the stage where they can provide input for surface wave generation.

A short cut to bridge the gap between the EGWW correlation model and the hydrocode is an intermediate code. Unlike the hydrocode which is a full-blown simulation of the actual state of the explosion, the intermediate code

considers only the phase after the shock energy is released. The emission of
the remaining internal energy from the explosion to the surrounding water
can be treated simply and easily. Since no thermodynamic exchange remains
significant in this stage, the phenomonology of this phase is described by the
potential flow theory where the fluid is assumed to be incompressible and
inviscid.

The technique used in the simulation is the Boundary Integral Equation
Method (BIEM). An axisymmetric formulation is used with time-stepping of
the Lagrangian points on the boundary surfaces. On the free surface, it is
done by time-integration of the nonlinear kinematic and dynamic boundary
conditions that apply on the boundary.

The method of time-stepping to follow an initial value problem using BIEM
was first proposed by Longuet-Higgins and Cokelet (1976) to study the evolution of steep waves. Underwater explosions were studied by Wilkerson (1988)
using BIEM, assuming constant potential and normal velocity in each element.
The model presented here takes linear elements and, within a segment, the potential and normal velocity are linearly related to their nodal values at the
end-points of each segment.

2. Mathematical Formulation

2.1. *Statement of the Problem*

An explosion is described by Cole (1948) as being a chemical reaction
characterized by the extreme rapidity of the process and the release of large
amounts of energy. The original material (usually solid) is converted into
a gaseous mixture at a very high temperature and pressure ($\sim 3000°C$ and
$\sim 50,000$ atm). The explosion converts the inherently unstable explosive material into a more stable product.

The explosion creates a hot mass of gas at a tremendous pressure. This
hot mass of gas, in turn, disturbs the surrounding medium. In underwater
explosions, this medium is water. The explosion's initial disturbance to the
water is caused by the arrival of the pressure wave from the explosion. The
fact that water is compressible causes pressure to be transmitted through the
fluid at a large, but finite velocity. The outward (away from the explosion)
movement of the water mass relieves the pressure wave. Once initiated, the
disturbance propagates as a shock wave. The gas sphere or bubble expands

radially at a decreasing rate due to the kinetic inertia of the fluid surrounding it. The gas bubble overexpands until the pressure inside becomes a fraction of the surrounding fluid pressure. At this stage, the outward motion stops and the bubble begins to contract at an increasing rate. While this happens, the pressure inside builds up until the inward motion stops. This dynamic behavior occurs even as the bubble constantly floats upwards towards the free surface due to buoyancy.

When the bubble hits the free surface, the pressure difference between the pressure inside the bubble and the atmospheric pressure above the free surface cannot be sustained for long. The bubble therefore explodes releasing high pressure gases into the air and causing a bowl-shaped crater. The disturbance caused by the crater collapse results in the propagation of very high amplitude waves away from the explosion.

The explosion bubble problem is similar to the cavitation bubble problem. Cavitation bubbles affect propeller and turbine blade performance and underwater acoustics. A number of investigations on the behavior of cavity bubbles have been conducted either experimentally or through computer simulations, for instance, Plesset and Chapman (1971), Chahine (1977, 1982, and 1990), Lauterborn and Bolle (1975), Shima and Nakajima (1977), Blake and Gibson (1981), Prosperetti (1982), and Dommermuth and Yue (1987). On the other hand, there are few reports on explosion-generated bubbles in the literature. The two works which are relevant to this topic are Wilkerson (1988) and Chahine *et al.* (1988).

The initial disturbance from the underwater explosion results in the propagation of a shock wave. This is followed by the coalescing of the hot reacting gases into a bubble of very high vapor pressure. To formulate and solve this problem, an incompressibility condition is imposed on the fluid mass and a time-simulation of the bubble-fluid system is carried out numerically. The brief period of the compressible phase during which the shock wave propagates is not considered.

A Reynolds number can be obtained by using the bubble radius and bubble radial velocity as the characteristic length and velocity scales, respectively. The Reynolds numbers calculated for cavitation bubbles are reported by Blake *et al.* (1986) to be the order of 10^4. For explosion bubbles, the Reynolds number is higher and remains high for the greater part of the bubble growth and collapse. Hence, viscous forces can be ignored in comparison to the inertia forces to assume the fluid inviscid. For an initially vortex-free flow field in an inviscid fluid, Kelvin's theorem of vorticity conservation allows the assumption of potential flow. With the assumption of incompressibility, numerical simulations can be carried out using an accurate and efficient technique called the Boundary Integral Equation Method (BIEM).

2.2. *The Integral Equation*

Define a cartesian coordinate system (x, y, z) where z is the vertical coordinate with $z = -h$ being the bottom and $z = 0$ being the undisturbed free surface. The assumption of irrotationality allows the introduction of a scalar velocity potential $\phi(x, y, z)$ such that the velocity vector $\mathbf{U}$ in the flow field is given as

$$\mathbf{U} = \nabla\phi. \tag{9.1}$$

Also, irrotationality is fulfilled as

$$\nabla \times \mathbf{U} = 0. \tag{9.2}$$

The conservation of mass in incompressible fluids is given by

$$\nabla \cdot \mathbf{U} = 0. \tag{9.3}$$

Combining Equations (9.1) and (9.3) results in the Laplace equation

$$\nabla^2\phi = 0. \tag{9.4}$$

Now, introduce a second scalar potential $\psi(x, y, z)$ for which Green's second identity can be written in the form

$$\int_v \phi\nabla^2\psi\,dv - \int_v \psi\nabla^2\phi\,dv = \int_s (\phi\nabla\psi \cdot \mathbf{n} - \psi\nabla\phi \cdot \mathbf{n})ds$$

$$= \int_s \left(\phi\frac{\partial\psi}{\partial n} - \psi\frac{\partial\phi}{\partial n}\right)ds \tag{9.5}$$

where v represents an integral over the fluid volume, s represents an integral over the surface of the fluid, and $\mathbf{n}$ is the unit normal vector on the surface pointing outwards from the fluid domain. From Equation (9.4) the second integral on the left-hand side of Equation (9.5) goes to zero, and the equation becomes

$$\int_v \phi\nabla^2\psi\,dv = \int_s \left(\phi\frac{\partial\psi}{\partial n} - \psi\frac{\partial\phi}{\partial n}\right)ds. \tag{9.6}$$

Let ψ be of the form $\frac{1}{R}$, where

$$R = [(x - x')^2 + (y - y')^2 + (z - z')^2]^{\frac{1}{2}} = |\mathbf{X} - \mathbf{X}'|. \qquad (9.7)$$

Here, the point $\mathbf{X}(x, y, z)$ is the reference point and $\mathbf{X}'(x', y', z')$ is the source point. It is known that $\frac{1}{R}$ is a valid solution of Poisson's equation that

$$\nabla^2 \left(\frac{1}{R} \right) = -4\pi\delta(R), \qquad (9.8)$$

where $\delta(R)$ is the Dirac delta function. From Equations (9.7) and (9.8), the volume integral in Equation (9.6) gives $-4\pi\phi(x, y, z)$. Hence, Equation (9.5) is reduced to the form

$$4\pi\phi(x, y, z) = - \int_s \left(\phi \frac{\partial}{\partial n} - \frac{\partial\phi}{\partial n} \right) \frac{1}{R} \, ds. \qquad (9.9)$$

Substituting a cylindrical coordinate system (r, θ, z) for the cartesian coordinates (x, y, z), one has $r = (x^2 + y^2)^{\frac{1}{2}}$, $\theta = \arctan(y/x)$, $ds = r d\theta d\ell$, where $d\ell$ represents a line element on the r-z plane. The unit normal vector $\mathbf{n}$ pointing outwards from the fluid domain has the direction cosines n_z and n_r, where n_r is positive radially outwards from the explosion and n_z is positive upwards. Also, $\mathbf{X} = \mathbf{X}(r, \theta, z)$ and $\mathbf{X}' = \mathbf{X}'(r', \theta', z')$. To include the variation in time t, Equation (9.9) can be written as

$$4\pi\phi(r, \theta, z, t) = - \int_s \left(\phi \frac{\partial}{\partial n} - \frac{\partial\phi}{\partial n} \right) \frac{r'}{R} \, d\theta' d\ell' \qquad (9.10)$$

in a cylindrical polar coordinate system, where

$$R = \{r^2(\mathbf{X}) + r^2(\mathbf{X}') - 2r(\mathbf{X})r(\mathbf{X}') \cos[\theta(\mathbf{X}) - \theta(\mathbf{X}')] + [z(\mathbf{X}) - z(\mathbf{X}')]^2\}^{\frac{1}{2}}. \qquad (9.11)$$

In the integral on the right hand side of Equation (9.10), only R is a function of θ. Separating the integral with respect to θ, Equation (9.10) becomes

$$4\pi\phi(r, \theta, z, t) = - \int_\ell r' \left(\phi \frac{\partial}{\partial n} - \frac{\partial\phi}{\partial n} \right) \int_0^{2\pi} \frac{d\theta'}{R} d\ell'. \qquad (9.12)$$

The integral over θ can be evaluated easily to remove all dependance on θ. As a result, the surface integral becomes a line integral and Equation (9.12) becomes

$$4\pi\phi(r, z, t) = -\int_\ell r'\left(\phi\frac{\partial}{\partial n} - \frac{\partial\phi}{\partial n}\right)\gamma(r, z;\, r', z')d\ell',\qquad (9.13)$$

where γ is the Rankine ring source function and

$$\gamma(r, z;\, r', z') = \int_0^{2\pi}\frac{d\theta'}{R} = \frac{4K(m)}{\rho_1}.\qquad (9.14)$$

Here, $K(m)$ is the complete elliptic integral of the first kind and the parameter m is defined as

$$m = 1 - \frac{\rho_0^2}{\rho_1^2},$$

where

$$\begin{aligned}
\rho_0^2 &= (z - z')^2 + (r - r')^2\\
\rho_1^2 &= (z - z')^2 + (r + r')^2\,.
\end{aligned}\qquad (9.15)$$

As shown by Newman (1985), the normal derivative of γ is given by

$$\begin{aligned}
\gamma_n &= \int_0^{2\pi}\frac{\partial}{\partial n}\left(\frac{d\theta'}{R}\right)\\
&= \frac{4(z - z')E(m)n_z}{\rho_0^2\rho_1} + \left[\frac{4(r - r')E(m)}{\rho_0^2\rho_1} + 2\frac{E(m) - K(m)}{r'\rho_1}\right]n_r,\qquad (9.16)
\end{aligned}$$

where $E(m)$ is the complete elliptic integral of the second kind; n_z and n_r have been defined as the direction cosines of the unit normal vector pointing outwards from the fluid domain. When $\rho_0 \to 0$, γ and γ_n develop logarithmic singularities.

As indicated in Equations (9.6)–(9.9), $-\int_v \nabla^2(\frac{1}{R})dv = 4\pi$, where 4π corresponds to the included solid angle at the point inside the fluid domain. On a smooth surface, of which the tangential plane is exposed to one half of the space; the included angle is 2π. If there is any discontinuity in the computational perimeter, the included angle has to be handled with care. In general, the included angle is the solid angle in space enclosed by the surrounding boundaries. For an axial symmetrical conic enclosure of half vortex angle θ, the included angle $\alpha = 2\pi(1 - \cos\theta)$. At the intersection of the matching surface and the initially calm free surface, the included angle is one quarter of the whole space, equal to π. Hence, Equation (9.13) can be written as

$$\alpha(r, z, t)\phi(r, z, t) = -\int_{\ell} r'\left(\phi\frac{\partial}{\partial n} - \frac{\partial\phi}{\partial n}\right)\gamma(r, z; r', z')d\ell', \qquad (9.17)$$

where $\alpha(r, z, t)$ generally denotes the included angle at the boundary point (r, z). In an axisymmetric flow field, only a radial section of the computational domain needs to be considered.

2.3. Boundary Conditions

The boundary conditions prescribed for the problem are

(a) Bottom Boundary Condition

For an impermeable rigid bottom, no flow occurs normal to the bottom surface. This is expressed mathematically as

$$\frac{\partial\phi}{\partial n} = 0, \qquad \text{at} \quad z = -h, \qquad (9.18)$$

which is easily satisfied by augmenting $\gamma(r, z; r', z')$ with its image at the bottom, $\gamma(r, z; r', -z' - 2h)$. By writing

$$G(r, z; r', z') = \gamma(r, z; r', z') + \gamma(r, z; r', -z' - 2h), \qquad (9.19)$$

and inserting G instead of γ in the field equation, Equation (9.17) becomes

$$\alpha(r, z)\phi(r, z) = -\int_{\ell} r'\left(\phi\frac{\partial}{\partial n} - \frac{\partial\phi}{\partial n}\right)G(r, z; r', z')d\ell'. \qquad (9.20)$$

It is understood that the field equation in the present form satisfies the bottom boundary condition identically.

The expressions for $\gamma(r, z; r', -z' - 2h)$ and $\gamma_n(r, z; r', -z' - 2h)$ are given as

$$\gamma(r, z; r', -z' - 2h) = \frac{4K(m_1)}{\rho_3}, \qquad (9.21)$$

$$\gamma_n(r, z; r', -z' - 2h) = -\frac{4(z + z' + 2h)E(m_1)n_z}{\rho_2^2\rho_3}$$

$$+ \left[\frac{4(r - r')E(m_1)}{\rho_2^2\rho_3} + 2\frac{E(m_1) - K(m_1)}{r'\rho_3}\right]n_r, \qquad (9.22)$$

where

$$m_1 = 1 - \frac{\rho_2^2}{\rho_3^2} \,,$$

$$\rho_2^2 = (z + z' + 2h)^2 + (r - r')^2 \,, \tag{9.23}$$

$$\rho_3^2 = (z + z' + 2h)^2 + (r + r')^2 \,.$$

Here $K(m_1)$ and $E(m_1)$ as defined earlier are the elliptic integral of the first and second kind, respectively.

(b) Bubble Boundary Condition

The kinematic boundary condition on the bubble is written as

$$\frac{D\mathbf{X}}{Dt} = \nabla\phi \,, \tag{9.24}$$

where $\frac{D}{Dt}$ denotes differentiation following a fluid particle or the substantial derivative. $\nabla\phi$ can be split into two components, $\frac{\partial\phi}{\partial n}$, which is normal to the fluid surface and $\frac{\partial\phi}{\partial s}$ which is tangential to the fluid surface. Also $\frac{D\mathbf{X}}{Dt}$ can be split into $\frac{Dr}{Dt}$ and $\frac{Dz}{Dt}$ that

$$\frac{Dr}{Dt} = \frac{\partial\phi}{\partial n}n_r + \frac{\partial\phi}{\partial s}n_z \,,$$

and

$$\frac{Dz}{Dt} = \frac{\partial\phi}{\partial n}n_z - \frac{\partial\phi}{\partial s}n_r \,. \tag{9.25}$$

The dynamic boundary condition on the bubble surface can be expressed by a relation between the vapor pressure inside the bubble p, and the pressure outside the bubble from Bernoulli's equation

$$p = p_a - \rho\frac{\partial\phi}{\partial t} - \frac{1}{2}\rho U^2 - \rho g z \,, \tag{9.26}$$

where p_a is a reference pressure (taken here equal to the atmospheric pressure), ρ is the fluid density, and g is the acceleration due to gravity. Rearranging and adding the term $\rho\mathbf{U} \cdot \nabla\phi$ to both sides of the equation, one has

$$\rho\left(\frac{\partial\phi}{\partial t} + \mathbf{U} \cdot \nabla\phi\right) = (p_a - p) + \rho\left(\mathbf{U} \cdot \nabla\phi - \frac{1}{2}U^2\right) - \rho g z \,. \tag{9.27}$$

By writing

$$\mathbf{U} \cdot \nabla \phi = U^2 = \left(\frac{\partial \phi}{\partial n}\right)^2 + \left(\frac{\partial \phi}{\partial s}\right)^2, \qquad (9.28)$$

and recognizing the left-hand side to be the substantial derivative of ϕ, the dynamic boundary condition on the bubble is given by

$$\frac{D\phi}{Dt} = \left(\frac{p_a - p}{\rho}\right) + \frac{1}{2}U^2 - gz. \qquad (9.29)$$

(c) Free-Surface Boundary Condition

The kinematic boundary condition on the free surface is the same as that specified for the bubble surface, Equation (9.25). The dynamic boundary condition on the free surface is a pressure relation. Since the pressure on the free surface is the atmospheric pressure, the dynamic boundary condition on the free surface is simply

$$\frac{D\phi}{Dt} = \frac{1}{2}U^2 - gz. \qquad (9.30)$$

(d) Far Field Condition

To obtain a mathematically unique solution, a far field condition has to be posed. In this initial value nonlinear problem, a far field closure can be imposed by matching the nonlinear potential inside the computational domain to a linear solution of transient radiated waves at a cylindrical matching boundary placed at a far distance from the origin (Wang and Nadiga, 1990). This is justified on the premise that, in a three-dimensional problem, the energy density of the radiated outgoing waves decreases rapidly with radial distance and the radiated waves become linear some distance away. However, this requires that the matching boundary be placed sufficiently far away from the origin so that the linear wave theory can be validly applied as the simulated time progesses.

To shorten the computing time, the matching boundary can be placed at a reasonably far location but still allow fluid to flow through the boundary freely. This is possible because the normal derviatives of the velocity potential are known from the solution at the previous time step. This information is then fed into the next time step as the initial condition. Let the normal velocity at the matching boundary $r = A$ be $V(A, z, t)$, then the boundary condition at the matching surface can be written as

$$\phi_r(r, z, t)|_{r=A} = V(A, z, t).$$ (9.31)

This manipulation is consistent with the boundary condition applied on the bubble surface. An error check with the conservation laws must be applied at each time step of integration to ensure that mass and energy are conserved.

Since the location of the radial boundary is held constant at $r = A$ at a fixed point (r, z) on the boundary, the condition requires

$$\frac{D\phi}{Dt} = \frac{\partial\phi}{\partial t} = -\frac{1}{2}V^2 - gz.$$ (9.32)

One may relax the vertical movement on the boundary, then the kinematic condition for a marker point (r, z) is given by

$$\frac{Dr}{Dt} = 0,$$

$$\frac{Dz}{Dt} = \frac{\partial\phi}{\partial r}n_z - \frac{\partial\phi}{\partial s}n_r.$$ (9.33)

The dynamic condition for this case is then

$$\frac{D\phi}{Dt} = -\frac{1}{2}\left(\frac{\partial\phi}{\partial r}\right)^2 + \frac{1}{2}\left(\frac{\partial\phi}{\partial s}\right)^2 - gz.$$ (9.34)

2.4. *Initial Conditions and Bubble Dynamics*

For a spherical bubble of a given radius and pressure, the radial velocity can be obtained by using Rayleigh's equation (1917). The conservation of mass flux requires that

$$4\pi r_0^2 u_0 = 4\pi r^2 u,$$ (9.35)

where r_0 is a reference radius of the bubble and u_0 is the radial velocity of the fluid mass at this radius while r and u are the instantaneous values of the bubble radius and bubble radial velocity, respectively. The energy released from the bubble can be calculated from the kinetic energy of the fluid mass outside the bubble which is given as

$$\frac{1}{2}\rho\int_{r_0}^{\infty}4\pi r^2 u^2 dr = 2\pi\rho u_0^2 r_0^3.$$ (9.36)

The work done by the bubble against a pressure difference between the ambient pressure p_a and the pressure inside the bubble p is given as

$$Q = \frac{4\pi}{3}(r_m^3 - r_0^3)(p_a - p),\qquad(9.37)$$

where r_m is the maximum radius to which the bubble expands. Equating the work done by the bubble to the kinetic energy of the fluid mass outside gives

$$u_0^2 = \frac{2\Delta p}{3\rho}\left(\frac{r_m^3}{r_0^3} - 1\right),\qquad(9.38)$$

where $\Delta p = p_a - p$. With respect to the reference radius r_0 and velocity u_0, the velocity of the bubble of radius r is given by

$$\left(\frac{dr}{dt}\right)^2 = \left(\frac{r_0}{r}\right)^3 u_0^2 + \frac{2}{3}\left(\frac{p - p_a}{\rho}\right)\left(1 - \frac{r_0^3}{r^3}\right).\qquad(9.39)$$

For the expansion of an explosion bubble, the input parameters are the yield and the depth of burst. The pressure inside the explosion bubble can be related to the explosion parameters by a pressure balance equation. For adiabatic expansion, Cole (1948) gives

$$p = 7.8\left(\frac{W}{\frac{4}{3}\pi r^3}\right)^{\gamma},\qquad(9.40)$$

where W = weight of the explosives in grams of TNT, $\gamma = 1.25$, r is given in centimeters, and p is given in kilobars.

The energy balance can be expressed by the following equation:

$$2\pi\rho r^3\left(\frac{dr}{dt}\right)^2 + \frac{1}{3}\rho\pi r^3\left(\frac{dz}{dt}\right)^2 + \frac{4}{3}\pi r^3 P_0 + E(r) = Y,\qquad(9.41)$$

where ρ is the fluid density, P_0 is the hydrostatic pressure at the depth of explosion h_0, $E(r)$ represents the internal energy of the explosion and Y is the total energy of the explosion. The first term is the kinetic energy in the surrounding fluid resulting from expansion. The second term represents the buoyancy effect and $\frac{dz}{dt}$ is the upward velocity due to buoyancy. The internal energy is a function of the bubble radius r and can be evaluated by

$$E(r) = \int_{v(r)}^{\infty} p\,dv,\qquad(9.42)$$

where v is the bubble volume equal to $\frac{4}{3}\pi r^3$ for a spherical bubble of radius r. Using the expansion equation (9.40), the internal energy is given by

$$E(r) = 7.8 \frac{W}{\gamma - 1} \left(\frac{W}{v(r)} \right)^{\gamma - 1} . \tag{9.43}$$

The total energy released by 1 gram of TNT is about 1060 calories. The quantity Y in Equation (9.41) represents the total energy left after the shock wave is emitted. Assuming 58.5 percent of the total energy is emitted in the shock wave (Cole 1948), Y is taken to be 440 calories per gram of explosives. Consequently, Equation (9.41) can be used to predict the initial conditions of radius and velocity for a given explosion.

3. Numerical Implementation

The equation to be solved is

$$\alpha(r, z, t)\phi(r, z, t) = - \int_L r' \left(\phi \frac{\partial}{\partial n} - \frac{\partial \phi}{\partial n} \right) G(r, z; r', z')d\ell , \tag{9.44}$$

where

$$G(r, z; r', z') = \gamma(r, z; r', z') + \gamma(r, z; r', -z' - 2h) . \tag{9.45}$$

The above equation is written so as to satisfy the bottom boundary condition as shown in Subsection 2.3. The path of integration L is the trace of the computational boundary and $L = B + F + M$, where B refers to the bubble perimeter, F to the free surface, and M to the matching boundary. To discretize the above equation, the trace of the computational boundary L is divided into small segments. Hence,

$$L = \sum_{j=1}^{N} \ell_j , \tag{9.46}$$

where N is the total number of segments and ℓ_j is the arc-length of the jth segment. Within a segment, ϕ is linearly related to the nodal values ϕ_j and ϕ_{j-1} and, similarly, its normal derivative ϕ_n is linearly related to $\phi_{n,j}$ and $\phi_{n,j-1}$. Thus, a system with infinite degrees of freedom is reduced to one with finite degrees of freedom. Then, Equation (9.44) can be expressed in terms of nodal values of ϕ and ϕ_n and reduced to a set of linear algebraic equations with ϕ_j or $\phi_{n,j}$ as the unknown variable. The solution is obtainable on solving the equation by direct Gaussian elimination.

3.1. *Spline Interpolation*

The solution of the field equation gives the velocity component ϕ_n normal to the fluid surface. By using a finite difference scheme, the tangential velocity component ϕ_s can be obtained from the known values of ϕ and the segment arc-lengths. The values of the velocity potential and the locations of the points on the moving boundaries are then updated using the kinematic and dynamic boundary conditions on these boundaries. It is clear that any errors in evaluating the direction angle or its cosines (n_z, n_r) on the moving boundaries will cause errors in the positioning of these points. This might cause irregularities in the profile and an eventual breakdown in the simulation. Thus, it is very important that the trace of the computational surface be represented very accurately with an absence of any sharp discontinuities.

To satisfy the above mentioned requirements, an interpolation of splines is taken through the points to ensure smoothness of the profile within segments. To have continuous slopes at nodal points, and to ensure further that the second derivatives are piecewise continuous, cubic splines are used. Then, ϕ_n and ϕ are functions of the spline parameter and are related to their nodal values.

Let ζ be the cubic spline parameter such that $1 \geq \zeta \geq 0$, then over the jth element one may write

$$\phi_j(\zeta) = \phi_{j-1} + \xi(\zeta)[\phi_j - \phi_{j-1}],$$
$$\phi_{n,j}(\zeta) = \phi_{n,j-1} + \xi(\zeta)[\phi_{n,j} - \phi_{n,j-1}],$$

(9.47)

where

$$\xi(\zeta) = \frac{s_j(\zeta)}{\ell_j},$$

(9.48)

with

$$s_j(\zeta) = \int_0^\zeta J_j(\zeta')d\zeta',$$

(9.49)

and

$$\ell_j = s_j(1.0).$$

(9.50)

Here ℓ_j is the arc-length of the jth segment, and

$$J_j(\zeta) = \left[\left(\frac{dr}{d\zeta}\right)^2 + \left(\frac{dz}{d\zeta}\right)^2\right]^{\frac{1}{2}}, \qquad (9.51)$$

is the Jacobian of $s(\zeta)$.

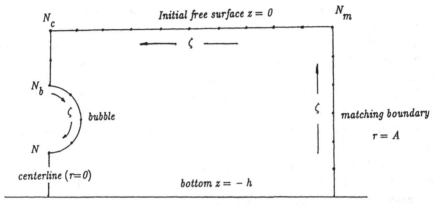

Fig. 9.1. Sketch showing discretization of the boundary over the computation domain.

The manner in which the perimeter is interpolated with cubic splines is shown in Figure 9.1. It is important to note that the assumption of axisymmetry reduces the computational surface to a line contour. Then, all parameters can be expressed as functions of the line element and the nodal values of the parameter. As shown in the figure, the point of intersection of the matching boundary with the free surface is N_m, while the intersection point of the free surface and the centerline is N_c. The point of intersection of the centerline and the top of the bubble is N_b, while the point of the intersection of the bottom of the bubble and the centerline is N. Also, the jth segment connects the points $j-1$ and j, so that

$$
\begin{aligned}
r_{j_{\zeta=0}} &= r_{j-1}, \\
r_{j_{\zeta=1}} &= r_j, \\
z_{j_{\zeta=0}} &= z_{j-1}, \\
z_{j_{\zeta=1}} &= z_j.
\end{aligned}
\qquad (9.52)
$$

The orientation of each segment is defined by

$$\theta_j = \tan^{-1}\left(\frac{z_j - z_{j-1}}{r_{j-1} - r_j}\right).$$ (9.53)

The continuity of the potential ϕ, the normal velocity ϕ_n and slopes n_z and n_r across the nodal points is expressed by

$$\lim_{\zeta \to 1}[\phi_j(1 - \zeta) - \phi_{j-1}(\zeta), \ \phi_{n,j}(1 - \zeta) - \phi_{n,j-1}(\zeta)] = 0,$$

$$\lim_{\zeta \to 1}[n_{z,j}(1 - \zeta) - n_{z,j-1}(\zeta), \ n_{r,j}(1 - \zeta) - n_{r,j-1}(\zeta)] = 0.$$

The interpolations along the various boundaries are specified by:

(a) Matching Boundary

$$r = A,$$
$$z = z_{j-1} + \zeta[z_j - z_{j-1}],$$ (9.54)
$$J_j = z_j - z_{j-1}.$$

(b) Free Surface

$$r = r_{j-1} + \zeta[r_j - r_{j-1}],$$
$$z = b_0(j) + \zeta b_1(j) + \zeta^2 b_2(j) + \zeta^3 b_3(j),$$ (9.55)
$$J_j = [(r_j - r_{j-1})^2 + (b_1(j) + 2\zeta b_2(j) + 3\zeta^2 b_3(j))^2]^{\frac{1}{2}},$$

where $b_\nu(j)$ are the spline coefficients of z on element j.

(c) Bubble Surface

If $|n_{z_j}| > |n_{r_j}|$,

$$r = r_{j-1} + \zeta[r_j - r_{j-1}],$$
$$z = b_0(j) + \zeta b_1(j) + \zeta^2 b_2(j) + \zeta^3 b_3(j),$$ (9.56)
$$J_j = [(r_j - r_{j-1})^2 + (b_1(j) + 2\zeta b_2(j) + 3\zeta^2 b_3(j))^2]^{\frac{1}{2}}.$$

If $|n_{z_j}| \leq |n_{r_j}|$,

$$r = a_0(j) + \zeta a_1(j) + \zeta^2 a_2(j) + \zeta^3 a_3(j) \,,$$

$$z = z_{j-1} + \zeta[z_j - z_{j-1}] \,, \tag{9.57}$$

$$J_j = [(z_j - z_{j-1})^2 + (a_1(j) + 2\zeta a_2(j) + 3\zeta^2 a_3(j))^2]^{\frac{1}{2}} \,.$$

Similarly, $a_\nu(j)$ are the spline coefficients of r on element j.

3.2. *Discretization of the Field Equation*

By writing the field equation in a discretized form, Equation (9.44) becomes

$$\alpha_i \phi_i = \sum_{j=1}^N \int_{j-1}^j r' \phi_n G d\ell'_j - \sum_{j=1}^N \int_{j-1}^j r' \phi G_n d\ell'_j \quad i = 0, 1, 2, \ldots, N \,. \tag{9.58}$$

It is understood that the subscript n denotes a normal derivative.

Now introduce the cubic spline parameter ζ into segment j. From Equations (9.49) and (9.50), it follows that

$$d\ell_j = ds_j(\zeta) = J_j(\zeta) d\zeta \,. \tag{9.59}$$

Incorporating the above in Equation (9.58) yields

$$\alpha_i \phi_i = \sum_{j=1}^N \int_0^1 r'(\zeta) J_j(\zeta) \phi_{n,j}(\zeta) G_{ij}(\zeta) d\zeta$$

$$- \sum_{j=1}^N \int_0^1 r'(\zeta) J_j(\zeta) \phi_j(\zeta) G_{n,ij}(\zeta) d\zeta \,, \tag{9.60}$$

where

$$G_{ij}(\zeta) = G(r_i, z_i; \ r'_j(\zeta), z'_j(\zeta)) \,,$$

and

$$G_{n,ij}(\zeta) = G_n(r_i, z_i; \ r'_j(\zeta), z'_j(\zeta)) \,. \tag{9.61}$$

Substituted by Equation (9.47), Equation (9.60) becomes

$$\alpha_i \phi_i = \sum_{j=1}^N \int_0^1 r'(\zeta) J_j(\zeta) G_{ij}(\zeta) [\phi_{n,j-1} + \xi(\zeta)(\phi_{n,j} - \phi_{n,j-1})] d\zeta$$

$$- \sum_{j=1}^N \int_0^1 r'(\zeta) J_j(\zeta) G_{n,ij}(\zeta) [\phi_{j-1} + \xi(\zeta)(\phi_j - \phi_{j-1})] d\zeta \,. \tag{9.62}$$

The nodal values ϕ_j, ϕ_{j-1}, $\phi_{n,j}$ and $\phi_{n,j-1}$ can be taken outside the integral sign to give

$$\alpha_i \phi_i = \sum_{j=1}^{N} \phi_{n,j} \int_0^1 r'(\zeta) J_j(\zeta) G_{ij}(\zeta) \xi(\zeta) d\zeta$$

$$+ \sum_{j=1}^{N} \phi_{n,j-1} \int_0^1 r'(\zeta) J_j(\zeta) G_{ij}(\zeta) [1 - \xi(\zeta)] d\zeta$$

$$- \sum_{j=1}^{N} \phi_j \int_0^1 r'(\zeta) J_j(\zeta) G_{n,ij}(\zeta) \xi(\zeta) d\zeta$$

$$- \sum_{j=1}^{N} \phi_{j-1} \int_0^1 r'(\zeta) J_j(\zeta) G_{n,ij}(\zeta) [1 - \xi(\zeta)] d\zeta . \qquad (9.63)$$

In general, the above equation can be written in the form of $N+1$ linear simultaneous equations:

$$\alpha_i \phi_i = \sum_{j=0}^{N} [H(i,j) \phi_j + Q(i,j) \phi_{n,j}] \quad i = 0, 1, 2, \ldots, N . \qquad (9.64)$$

where

$$H(i,j) = - \int_0^1 r'(\zeta) J_j(\zeta) G_{n,ij}(\zeta) \xi(\zeta) d\zeta$$

$$- \int_0^1 r'(\zeta) J_{j+1}(\zeta) G_{n,ij+1}(\zeta) [1 - \xi(\zeta)] d\zeta \quad 0 < j < N$$

$$H(i,0) = - \int_0^1 r'(\zeta) J_1(\zeta) G_{n,i1}(\zeta) [1 - \xi(\zeta)] d\zeta$$

$$H(i,N) = - \int_0^1 r'(\zeta) J_N(\zeta) G_{n,iN}(\zeta) \xi(\zeta) d\zeta , \qquad (9.65)$$

and

$$Q(i, j) = \int_0^1 r'(\zeta) J_j(\zeta) G_{ij}(\zeta) \xi(\zeta) d\zeta$$

$$+ \int_0^1 r'(\zeta) J_{j+1}(\zeta) G_{ij+1}(\zeta) [1 - \xi(\zeta)] d\zeta \quad 0 < j < N$$

$$Q(i, 0) = \int_0^1 r'(\zeta) J_1(\zeta) G_{i1}(\zeta) [1 - \xi(\zeta)] d\zeta$$

$$Q(i, N) = \int_0^1 r'(\zeta) J_N(\zeta) G_{iN}(\zeta) \xi(\zeta) d\zeta . \tag{9.66}$$

The integrals in the functions H and Q can be evaluated using Gaussian quadratures. However, when $i = j$ or $i = j + 1$, G_{ij} and $\frac{\partial G_{ij}}{\partial n}$ develop logarithmic singularities. The integration of the singular terms can be performed by making use of the quadrature for logarithmic singularities.

4. Method of Solution

The solution of the problem is governed by the integral equation (9.44), of which the discretized form is given by (9.64). The boundary is discretized into N segments, on which $N + 1$ nodal values of ϕ or ϕ_n are sought as a function of time t. With ϕ and ϕ_n being known at any instant, the tangential derivatives ϕ_s at the nodal points can also be determined for that particular instant. It follows then through time integration to obtain ϕ, ϕ_n and ϕ_s for the following instant at $t + \Delta t$. In general, the solution process consists of two steps, an Eulerian phase to solve the field equation at a fixed time t and a Lagrangian scheme to advance the solution to the next time step.

4.1. *Solution of the Field Equation*

Equation (9.64) is a linear algebraic system corresponding to the integral equation (9.44). As shown in (9.44), it is a mixture of Fredholm integral equations of the first and second kind. At a nodal point, either ϕ or ϕ_n is known but not both. To initiate the computation, the free surface is assumed to be calm and the potential ϕ is constant in the fluid domain except adjacent to the small spherical surface of the bubble at the initial expansion. This corresponds to a very short instant after the explosion while the disturbance has not reached the ambient fluid. In particular, the fluid on the free surface and the matching boundary is assumed to be undisturbed at the initial instant, and the potential ϕ is arbitrarily specified to be zero. On the bubble surface, the potential ϕ is unknown, but the radial velocity ϕ_n can be determined. This allows direct Gaussian elimination on the matrix equation (9.64) to solve on the bubble surface and the normal velocity ϕ_n on the free surface and the matching boundary.

After the initiation, the value of ϕ is specified at every nodal point so that Equation (9.64) can be rewritten by grouping all the unknowns on the left-hand side and all the known quantities on the right in matrix notation,

$$A_{ij}\phi_{n,j} = B_i, \quad i = 0, 1, 2, \ldots, N. \tag{9.67}$$

where A_{ij} is an $(N+1) \times (N+1)$ square matrix of coefficients equal to $Q(i, j)$, and B_i is an $(N+1)$ element column matrix of coefficients equal to $\alpha_i \phi_i - \sum_{j=0}^{N} H(i, j)\phi_j$.

Again, this equation can be solved by direct Gaussian elimination for the values of ϕ_n at the $N+1$ collocation points.

4.2. *Time-Stepping*

By following Lagrangian points on the bubble surface and on the free surface, one has at a given time instant the position and the velocity potential at all the points on the moving boundaries. To perform time-stepping, one has to integrate the dynamic and kinematic boundary conditions with respect to time to obtain the new positions of Lagrangian points on the bubble and on the free surface.

In order to integrate the dynamic and kinematic boundary conditions, the velocity vector (ϕ_n, ϕ_s) has to be specified completely. In solving the field equation, ϕ_n is known on all boundaries after the first time step. However, the velocity component tangential to the boundary does not come out explicitly in solving the field equation. By using a finite difference scheme, the velocity component ϕ_s can be obtained.

All the boundary segments are approximated by cubic splines and can be expressed as a function of the cubic spline parameter ζ. Accordingly, the potential ϕ is a function of ζ as was shown in Equation (9.47) and the tangential derivative $\partial \phi / \partial \zeta$ can be calculated in a straightforward fashion. From Equation (9.49), it is clear that $\frac{ds}{d\zeta} = J_j$. Hence,

$$\phi_{s,j} = \frac{\partial \phi_j}{\partial s} = \frac{\partial \phi_j}{\partial \zeta} \bigg/ \frac{\partial s_j}{\partial \zeta} = \frac{1}{J_j} \frac{\partial \phi_j}{\partial \zeta}. \tag{9.68}$$

The kinematic and dynamic boundary conditions can now be integrated over time to give the new potential and the new position of points on the boundary surfaces. The governing differential equation for the bubble surface was specified by Equations (9.25) and (9.29), the free surface by Equations (9.25) and (9.30), and the matching surface by Equations (9.33) and (9.34). These equations are first-order ordinary differential equations in time t. To integrate, a multi-step scheme of the Adam-Bashforth-Moulton method is appropriate. It is generally the predictor-corrector method of Milne (1953). For an equation of the form

$$\frac{dy}{dt} = f(y, t),\qquad\qquad(9.69)$$

the stepping process is as follows:

(1) the predicted value of y at time $t_1 = t_0 + \Delta t$

$$y_{1p} = y_0 + \frac{\Delta t}{24}(55f_0 - 59f_{-1} + 37f_{-2} - 9f_{-3}),\qquad(9.70)$$

(2) the corrected solution of y at t_1

$$y_{1c} = y_0 + \frac{\Delta t}{24}(9f_{1p} + 19f_0 - 5f_{-1} + f_{-2}).\qquad(9.71)$$

It is a two-step integration from a third-order polynomial fit to the derivatives of the latest four solution points. Since the method requires information from three previous time steps, a simple Euler's method is used for the first three time-steps from the initial condition.

The integration of the kinematic and dynamic boundary conditions constitutes the time-stepping of the boundary position and the velocity potential. Then, the known values of the potential ϕ on the bubble, on the free surface, and on the matching boundary can be substituted in the field equation to advance for solutions of the next time step.

5. Simulation of Bubble Breaking

5.1. *Initiation of Bubble Breaking*

When the gaseous mixture expands and breaches the free surface leading to bubble breaking, the phenomenon becomes complicated to handle numerically.

In an underwater explosion, the submerged detonation instantly produces a spherical volume of hot gas. High temperatures and pressures then result in two distinctive disturbances in the surrounding fluid; one, the emission of a shock wave traveling outward, and two, a radial motion of the fluid resulting from the expansion of the hot gas and explosive debris. Immediately followed then is the growth of a spray dome right above the explosion. Water under this dome rises rapidly to form a vertical plume. The maximum height of the water column is governed by the initial pressure and velocity which in turn depend on the charge weight and the burst depth below the surface. For sufficiently deep detonations, the dome and the subsequent plume may not be visible at all. For shallow charges, two types of plume spray can be formed; it can be a relatively narrow column of very high spray or a radial plume projected outward in all directions. Cole (1948) shows that the types of the plume can be correlated with the period of the bubble oscillation and the time at which the plume leaves the water surface, which again correspond to the charge weight and depth. In any case, this plume is a mixture of water spray and explosion products. Eventually, opposed by the force of gravity, capillarity and air resistance, the plume breaks into water droplets together with explosion debris falling back on the water surface. While these fallouts of the plume spray may generate a lot of turbidity on the water surface, it is the cavity formed after venting of the explosive debris that is responsible for the surface wave generation.

Fluid motions from near surface explosions of small HE charges have been reported by a number of researchers. As a function of the charge weight and depth, the process of dome, plume and cavity formation was schematically demonstrated in Figure 1.2 by Kedrinskii (1978). In particular, it showed that in some cases, a very high upward moving jet rose from the dome right after the beginning of the collapse of the gaseous bubble. Similar phenomena were observed in the growth and collapse of cavitation bubbles close to a free surface by Chahine (1977) and Blake and Gibson (1981).

The phenomenon of jet development was investigated theoretically by Longuet-Higgins (1983). It was shown that as the surface dome rises during the process of bubble expansion, the vortex angle progressively decreases while the surface dome becomes more peaked. As the vortex angle of the dome reduces to a certain critical angle, the motion starts a jolt, such that the jet velocity and acceleration both become infinite. Longuet-Higgins demonstrated that the critical angle is $2 \arctan 2^{1/2}$ by using a gravity-free potential flow model of

Dirichlet's hyperboloid. The comformity of Dirichlet hyperbolic model to the profile of two-dimensional breaking waves (Longuet-Higgins, 1972) was confirmed numerically by McIver and Peregrine (1981).

Physically, the jet or plume arises from instability of the free surface when the vortex of the dome surface reaches its critical angle. Numerically, this is the instant that the computation breaks down, as the velocity and acceleration are becoming infinite. To continue the computation, it is assumed that the bubble is simply broken up at the apex at this very moment, ignoring the jet formation. The fallouts from jet or plume may indeed produce turbidity on the water surface but is assumed to have no effect on wave disturbances.

5.2. *Bubble and Free-Surface Intersection Point*

As soon as the bubble breaks, there arises another difficult problem in the numerical scheme. This is because there are two different kinematic boundary conditions imposed at the intersection point. There are two different ϕ_n values because of two different directional normals existing at the sharp corner of the intersection point. This results in a singularity at the intersection point. This singularity can have a global effect on the simulation and, if not properly treated, can cause a breakdown of the simulation. To circumvent this problem, Lin (1984) proposed that both ϕ and ϕ_n at the intersection point be treated as known. In effect, this reduces the number of unknowns by one.

In a different approach, the unknown is assumed to remain single valued at each collocation point including the sharp corner at the intersection point. It is understood that both ϕ_n and ϕ_s are continuous at any collocation point except the intersection corner. Hence, it is further postulated that these derivatives remain single valued at the corner, although the rates of change of these values are drastically distinctive around the corner as compared to that at other points.

Following the convention adopted in the numerical scheme, the value ϕ_s at point N_c is defined by a finite difference in the form

$$\phi_{s,N_c} = \frac{\phi_{N_c} - \phi_{N_c-1}}{\ell_{N_c}}, \tag{9.72}$$

where ℓ_{N_c} is the arc length of the segment N_c, between points $N_c - 1$ and N_c. On the other hand, it possesses a sharply different tangential derivative at this point, both in magnitude and direction, if interpolated from segment $N_c + 1$,

$$\phi_{s,N_c+1} = \frac{\phi_{N_c+1} - \phi_{N_c}}{\ell_{N_c+1}}. \tag{9.73}$$

The orientations of segments N_c and $N_c + 1$ are known from spline interpolation (9.53) and is given as θ_{N_c} and $\theta_{N_{c+1}}$, respectively, then the angle between the segments N_c and $N_c + 1$ is obtained by

$$\beta = \pi + \theta_{N_{c+1}} - \theta_{N_c}. \tag{9.74}$$

Consequently, the magnitude of the tangential derivative at the corner can be justified by taking one half of the vector summation of that on the two sides of the corner[a]

$$\phi_{s,N_c^*} = \pm \frac{1}{2}[\phi_{s,N_c}^2 + \phi_{s,N_c+1}^2 - 2\phi_{s,N_c}\phi_{s,N_c+1}\cos\beta]^{1/2}. \tag{9.75}$$

Here ϕ_{s,N_c^*} takes the + sign when ϕ_{s,N_c} is positive and $-$ sign when ϕ_{s,N_c} is negative, and the tangential direction is supposed to be

$$\theta_{N_c^*} = \theta_{N_c} - \gamma, \tag{9.76}$$

where

$$\gamma = \pm \cos^{-1}\frac{\phi_{s,N_c^*}^2 + \phi_{s,N_c}^2 - \phi_{s,N_c+1}^2}{2\phi_{s,N_c^*}\phi_{s,N_c}}, \qquad \frac{\phi_{s,N_c}}{\phi_{s,N_c+1}} \gtrless 0. \tag{9.77}$$

The directional normal at the corner must be perpendicular to the tangential orientation and can be evaluated by

$$(n_r, n_z)_{N_c^*} = (\sin\theta_{N_c^*}, \cos\theta_{N_c^*}). \tag{9.78}$$

By considering ϕ_n and ϕ_s at the breaking corner to be continuous and single valued, the included angle α in the field equation (9.44) also needs to be adjusted at the corner point (r_{N_c}, z_{N_c}); that is simply

$$\alpha(r_{N_c}, z_{N_c}, t) = 2\beta. \tag{9.79}$$

---

[a]It is noted that the tangential derivative defined this way is equivalent to a second-order finite difference evaluated at collocation point j, i.e.,

$$\phi_{s,j} = \frac{1}{2}\left[\frac{\phi_{j+1} - \phi_j}{\ell_{j+1}} + \frac{\phi_j - \phi_{j-1}}{\ell_j}\right]$$

The directional normal at point N_c therefore is modified by substituting $(n_r, n_z)_{N_c^*}$ for $(n_r, n_z)_{N_c}$. Similarly, the analysis can be considered for the intersection point N_b at the time of breaking and the directional normal at N_b is substituted by $(n_r, n_z)_{N_b^*}$. Thus, the multivalued problem is approximated by a single value problem to solve ϕ_n at the corners N_c and N_b as well as all the other collocation points uniquely.

6. Parametric Presentation

6.1. *Sample of Typical Simulations*

To demonstrate the validity of the BIEM formulation, the expansion and contraction of cavity bubbles were simulated and are shown here. Presented as an example is the computation of a TNT explosion in water of 58 feet deep. The charge weight W is taken to be 100 pounds and the depth of burst (DOB) z_0 is 11.6 feet below the surface.

Figure 9.2 shows the numerically simulated bubble contours at 18 equal time intervals after the initiation. Sixty nodal points in total are used over the computational boundary. Referring to Figure 9.1, $N_m = 10$, $N_c = 40$, $N_b = 41$ and $N = 59$. The initial radius of the bubble is taken to be 6.5 feet, which is approximately 25% of the maximum attainable radius for this case. The time interval used in the presentation is 22.5 milliseconds. The time at the end of the simulation corresponds to 405 milliseconds elapsed after the initiation.

Figure 9.3 shows the vertical velocity variations along the centerline axis above and below the bubble. Again, it presents the same 18 time-steps. These vertical velocities are computed at five equally spaced points above and below the bubble surface. Straight lines are shown in the figure to connect the values at the bubble apex above and at bottom below. Only centerline velocities below the bubble are shown after breaking. The magnitude of the centerline velocity v presented in the figure is normalized by a characteristic velocity U equal to the square root of gz_0. The linear dimension of the vertical axis is normalized by the charge depth z_0. The position -1 on the vertical axis is the charge position.

The pressure variations above and below the bubble are shown in Figure 9.4. The centerline pressure is presented in a nondimensional form of $(p-p_a)/pgz_0$. The gas pressure inside the bubble is shown by a vertical straight line at any instant. The pressure above the bubble varies from the bubble gas

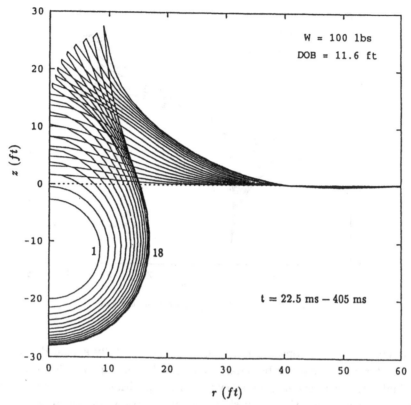

Fig. 9.2. Simulation of crater formation and surface movement for subsurface explosion in 58-feet deep water.

pressure at the top of the bubble surface to the atmospheric pressure at the free surface. The bubble gas pressure decreases with time and becomes equalized with the atmospheric pressure after the surface layer is broken. Only nine time-steps before breaking are shown in the figure. The pressure below the bubble is dominated by hydrostatics as indicated by the roughly linear trend on the pressure curve.

6.2. *Deep Water Simulations*

Simulations have been conducted for various yields from 200 pounds to 1 KT. For each yield, a number of charge submergence ratios are taken.

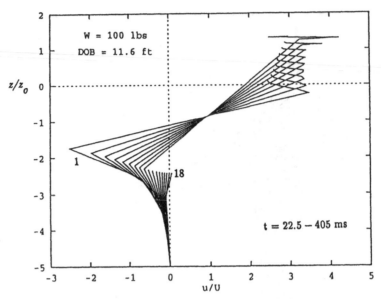

Fig. 9.3. Centerline velocity near the bubble.

Using BIEM, the computation boundary must be fully specified. For the deep water case, the lower boundary is specified to be equal to 2.5 times the attainable radius of the bubble below the charge position. In order to enhance the presentation, only the crater shape at the instant when the top surface of the bubble reaches its peak level is shown for each case. This crater form may then be taken as the input to the wave propagation model to determine the wave information at the far field. For the case of deeper charge submergence, the top surface may not even break in the first cycle of bubble expansion, but the peak surface height of the first cycle generally governs the propagation and the magnitude of the waves.

Graphic presentations for yields of 200 pounds, 1 ton, 10 tons, 100 tons, and 1 KT are shown in Figures 9.5–9.9, respectively. For each yield, six charge positions or depth of burst $(z/W)^{1/3}$ are shown in the figure. The linear scales of these figures are in feet.

6.3. *Shallow Water Simulations*

Similar to the deep water case, simulations for shallow water explosions

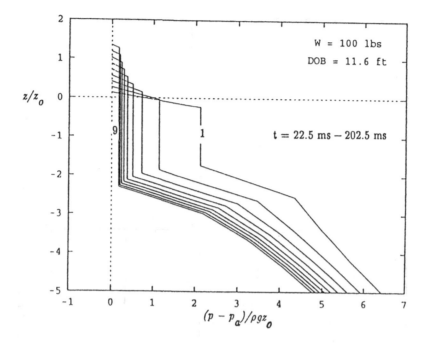

Fig. 9.4. Centerline fluid pressure before bubble breaking.

are conducted for five yields ranging from 200 pounds to 1 KT. Graphic presentations of the crater forms for the case of $W = 200$ pounds are shown in Figure 9.10. Again, the linear scales are in feet.

For each yield, five different depths are selected to demonstrate the effect of water depth on wave generation. The depth parameter is $d/W^{1/3}$. Figure 9.10 covers five subsets corresponding to $d/W^{1/3} = 2$, 3, 4, 5, and 6. For each $d/W^{1/3}$, four different charge positions are shown in the figure. The charge position parameter is taken to be the charge submergence to water depth ratio z/d.

Note that the product of the two parameters $d/W^{1/3}$ and z/d gives the depth of burst (DOB) ratio $z/W^{1/3}$, which has been defined to parameterize the charge position for the deep water cases. Consequently, the shallow water results provide also, although not directly, indications due to $z/W^{1/3}$ if compared with the deep water cases.

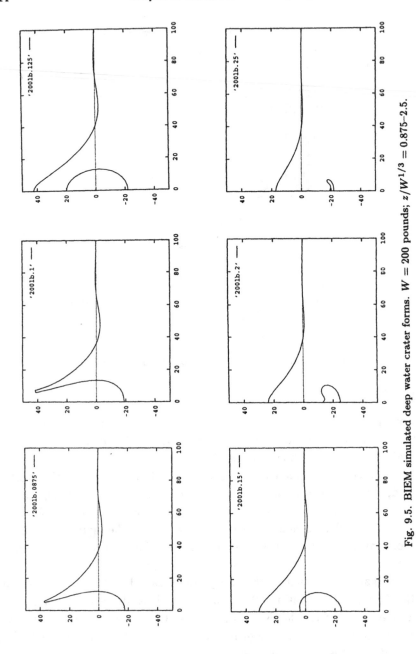

Fig. 9.5. BIEM simulated deep water crater forms. $W = 200$ pounds; $z/W^{1/3} = 0.875$–2.5.

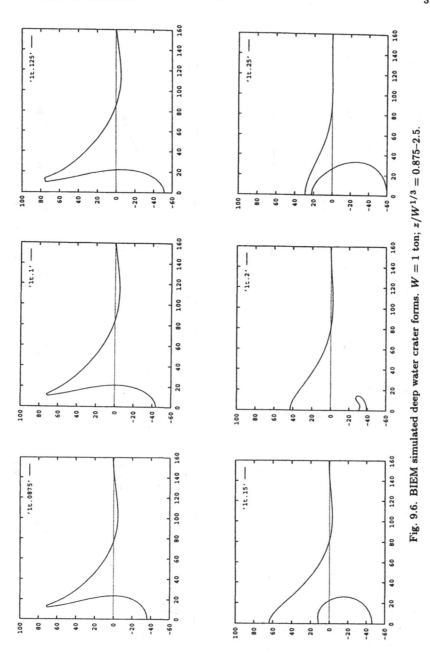

Fig. 9.6. BIEM simulated deep water crater forms. $W = 1$ ton; $z/W^{1/3} = 0.875$–2.5.

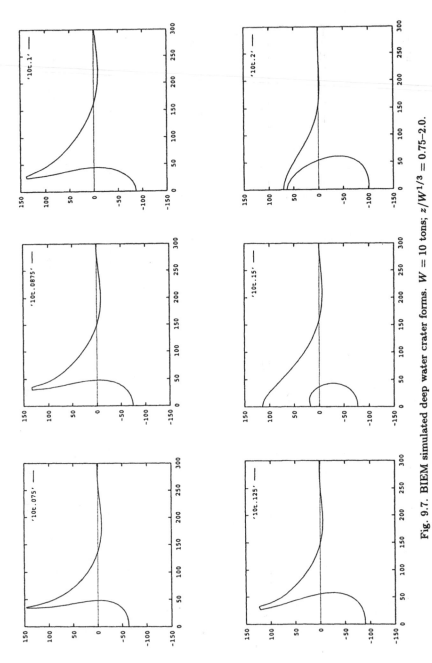

Fig. 9.7. BIEM simulated deep water crater forms. $W = 10$ tons; $z/W^{1/3} = 0.75$–2.0.

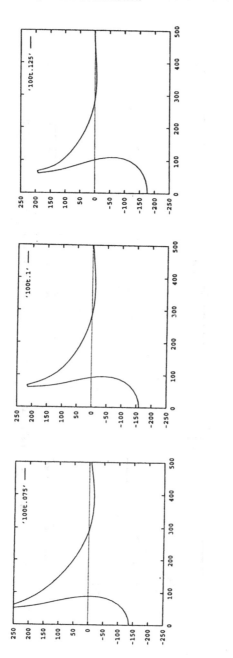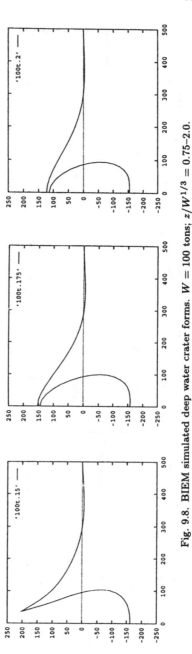

Fig. 9.8. BIEM simulated deep water crater forms. $W = 100$ tons; $z/W^{1/3} = 0.75$–2.0.

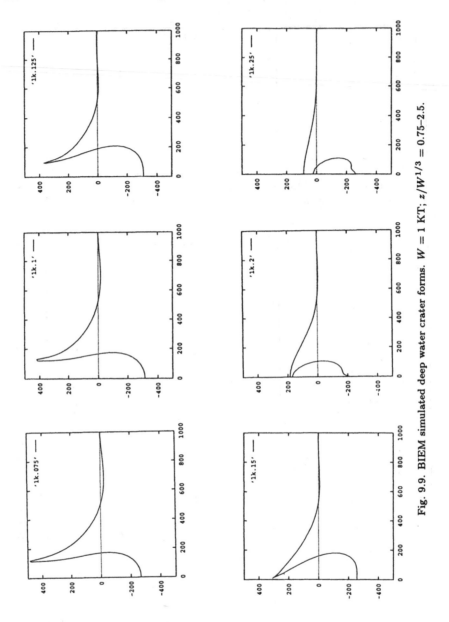

Fig. 9.9. BIEM simulated deep water crater forms. $W = 1$ KT; $z/W^{1/3} = 0.75$–2.5.

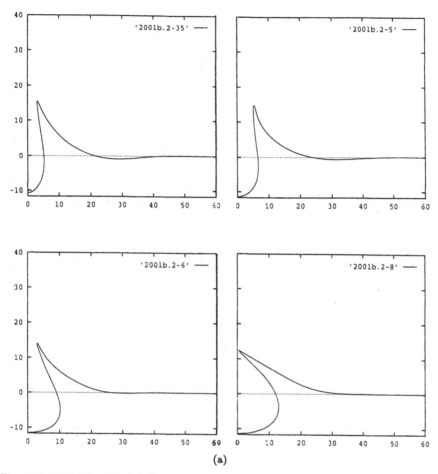

(a)

Fig. 9.10. BIEM simulated shallow water crater forms, $W = 200$ pounds. (a) $d/W^{1/3} = 2$; $z/d = 0.35$–0.8. (b) $d/W^{1/3} = 3$; $z/d = 0.25$–0.8. (c) $d/W^{1/3} = 4$; $z/d = 0.2$–0.8. (d) $d/W^{1/3} = 5$; $z/d = 0.2$–0.8. (e) $d/W^{1/3} = 6$; $z/d = 0.15$–0.8.

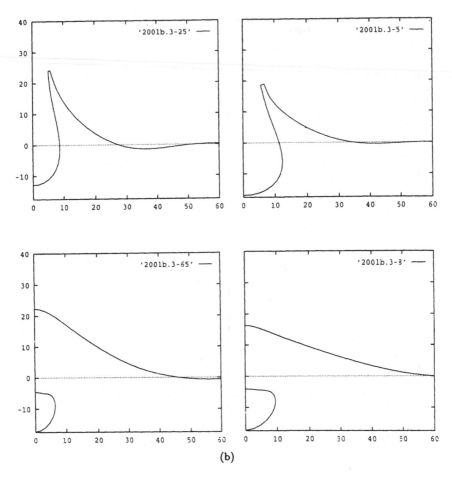

(b)

Fig. 9.10. (*Continued*)

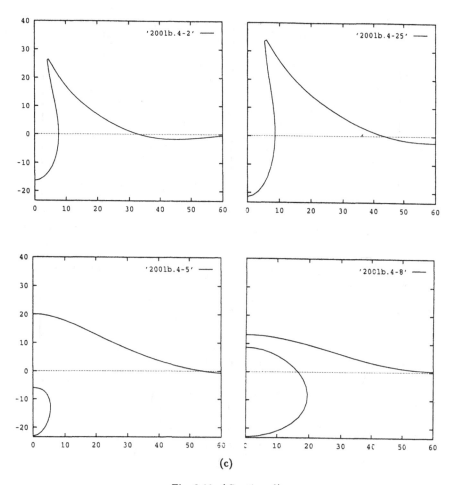

(c)

Fig. 9.10. (*Continued*)

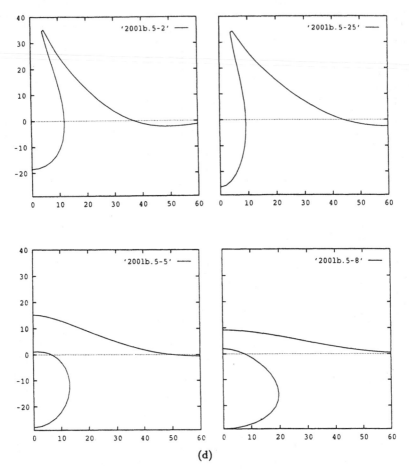

(d)

Fig. 9.10. (*Continued*)

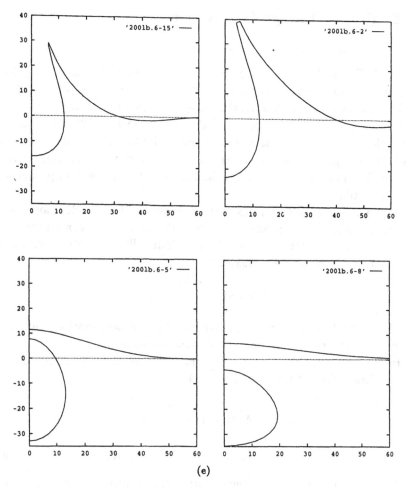

(e)

Fig. 9.10. (*Continued*)

7. Equivalent-Energy Correlations

7.1. *Equivalent-Energy Crater Forms*

The BIEM model provides a crater form, which would be directly applicable as input information for wave predictions if an appropriate wave prediction model were available. Currently, the wave height prediction for the deep water case is based on a linear Kranzer-Keller-type model (1959) approximation, where the crater form is assumed mathematically to have a Hankel transform (see Chapter 2). In order to connect the BIEM crater with the Kranzer-Keller wave model, a method to transform the BIEM crater into an energy equivalent crater is outlined here.

At the moment the crater lips reach their peaks, the particle velocities over the crater surface are approximately zero. The crater then can be considered to possess only potential energy. For a cylindrically symmetrical crater, the potential energy E can be evaluated by

$$E = \pi \rho g \int_0^\infty \eta^2(r) r \, dr \,, \tag{9.80}$$

where ρ is the density of water, g is the gravitational acceleration, and η is the crater elevation which is a function of radial distance r.

In particular, we shall assume the crater model to be of a parabolic zero volume form (Table 2.1, case 3) for which

$$\eta(r) = A\left[2\left(\frac{r}{R}\right)^2 - 1\right] \,, \quad r \leq R \,,$$

$$= 0 \,, \qquad\qquad r > R \,. \tag{9.81}$$

Here, R is the radius at which the elevation of the crater has the highest value A (see figure in Table 2.1) or the crater height. From Table 2.1, the potential energy for a parabolic zero volume crater is known to be $\pi \rho g A^2 R^2 / 6$.

To resemble the surface profile of the deep charge submergence case, we select the crater model to be of the hyperbolic form (Table 2.1, case 5), which is described by

$$\eta(r) = \frac{AR^3}{(R^2 + r^2)^{3/2}} \,. \tag{9.82}$$

The potential energy for this form is known to be $\pi \rho g A^2 R^2 / 4$ as shown in Table 2.1.

The potential energy for an BIEM crater is computed using Equation (9.80) and the equivalent crater height A is determined from the BIEM result. For consistency, the value of A is taken to be one half the total height from the bottom of the cavity to the tip of the lip. The value R is evaluated in accordance with Equation (9.81) if the surface is broken. Otherwise, Equation (9.82) is used to evaluate A and R.

7.2. *Summary of Correlated Results*

(a) *Deep Water Correlations*

Through the energy-equivalent method, the equivalent A and R for each simulated case are summarized in Table 9.1. The results are arranged for yields from 200 pounds to 1 KT, with the charge submergence as the parameter. One may compare these results with that presented in Equation (3.2) correlated by Le Méhauté (1970).

Equation (3.2a) is for the upper critical depths and Equation (3.2b) is for the lower critical depth. The BIEM results should compare only with Equation (3.2b), the lower critical depth case. Equation (3.2b) indicated that A is proportional to $W^{0.24}$ and R to $W^{0.3}$. The results shown in Table 9.1 generally support a $W^{1/4}$ law for both A and R.

It is noted that Table 9.1 includes results obtained by two model correlations (9.81) and (9.82). Figure 9.11 shows results of crater radius R as a function of yield W for those cases correlated using Equation (9.81). First, one may notice that the crater radius is not sensitive to the parameter of charge submergence $z/W^{1/3}$. Second, the radius R follows a $W^{1/4}$ correlation very closely. Designating R as the correlated value of R for the deep water case, its correlation with yield W can be approximated as follows:

$$R_0 = 8.97 \, W^{1/4} \, . \qquad (9.83)$$

To accept that the crater radius is independent of charge depth and converting Table 9.1 results in accordance with a single model (9.81), Table 9.2 is generated. In Table 9.2, the R values are taken according to Equation (9.83) to be R_0. The A values are adjusted and reevaluated in accordance with the concept of equivalent energy and presented as A_0.

Table 9.2 shows that while R_0 is independent of charge submergence z, A_0 varies as a function of $z/W^{1/3}$. Figure 9.12 further indicates that at a constant $z/W^{1/3}$, A_0 varies as $W^{1/4}$. The results provide a correlation of the following

Table 9.1. Energy equivalent correlation of crater forms (deep water).

Yield	$z/W^{1/3}$	A(ft)	R(ft)	Corr. Equ.
200 lbs	.875	27	29	9.81
	1	30	29	9.81
	1.25	43	21	9.82
	1.5	31	22	9.82
	2	24	19	9.82
	2.5	17	20	9.82
1 ton	.875	59	59	9.81
	1	58	63	9.81
	1.25	63	63	9.81
	1.5	64	41	9.82
	2	43	42	9.82
	2.5	29	46	9.82
10 ton	.75	105	103	9.81
	.875	104	115	9.81
	1	112	120	9.81
	1.25	106	110	9.81
	1.5	114	83	9.82
	2	71	85	9.82
100 ton	.75	200	222	9.81
	1	187	199	9.81
	1.25	184	203	9.81
	1.5	181	198	9.81
	1.75	153	155	9.82
	2	123	161	9.82
1 KT	.75	376	360	9.81
	1	365	354	9.81
	1.25	336	365	9.81
	1.5	278	366	9.81
	2	186	292	9.82
	2.5	87	313	9.82

Table 9.2. Energy equivalent crater forms adjusted in
accordance with Equations (9.81) and (9.83) (deep water).

Yield	$z/W^{1/3}$	A_0(ft)	R_0(ft)
200 lbs	.875	23	34
	1	26	34
	1.25	22	34
	1.5	16	34
	2	11	34
	2.5	8	34
1 ton	.875	58	60
	1	61	60
	1.25	66	60
	1.5	36	60
	2	25	60
	2.5	18	60
10 ton	.75	101	107
	.875	112	107
	1	126	107
	1.25	109	107
	1.5	72	107
	2	46	107
100 ton	.75	234	190
	1	196	190
	1.25	197	190
	1.5	189	190
	1.75	102	190
	2	85	190
1 KT	.75	402	337
	1	383	337
	1.25	364	337
	1.5	302	337
	2	132	337
	2.5	66	337

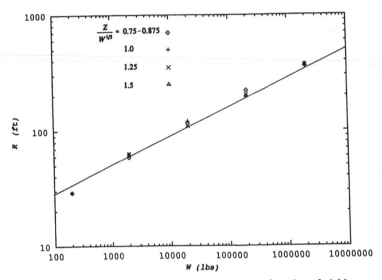

Fig. 9.11. Deep water crater radius plotted as a function of yield.

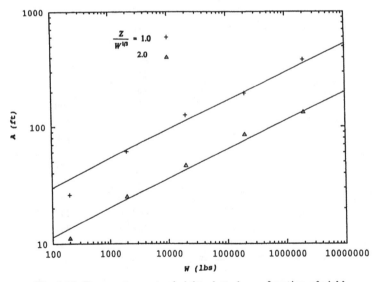

Fig. 9.12. Deep water crater height plotted as a function of yield.

form:

$$A_0 = 9.46\, fW^{1/4}\,, \tag{9.84}$$

where f is designated as a correction factor due to charge position z and is approximately given by

$$f = \exp\left(1 - \frac{z}{W^{1/3}}\right), \quad 0.75 < \frac{z}{W^{1/3}} < 2.5\,. \tag{9.85}$$

(b) *Shallow Water Correlations*

Following the same method applied to deep water correlations, the equivalent A and R is tabulated as a function of $d/W^{1/3}$ and z/d. Again, two correlation models (9.81) and (9.82) are used to obtain the results of A and R (Table 9.3).

Table 9.3. Energy equivalent correlation of crater forms (shallow water).

	$d/W^{1/3}$	z/d	A(ft)	R(ft)	Corr. Equ.
$W = 200$ lbs	2	.35	13	14	9.81
		.5	13	15	9.81
		.6	13	18	9.81
		.8	12	23	9.82
	3	.25	19	21	9.81
		.5	18	24	9.81
		.65	22	22	9.82
		.8	16	29	9.82
	4	.2	21	24	9.81
		.25	28	32	9.81
		.5	20	31	9.82
		.8	13	35	9.82
	5	.2	27	29	9.81
		.25	30	31	9.81
		.5	15	27	9.82
		.8	9	37	9.82
	6	.15	22	24	9.81
		.2	31	31	9.81
		.5	11	28	9.82
		.8	6	37	9.82

Table 9.3. (*Continued*)

	$d/W^{1/3}$	z/d	A(ft)	R(ft)	Corr. Equ.
$W = 1$ ton	2	.4	29	29	9.81
		.5	26	35	9.81
		.65	32	24	9.82
		.8	33	28	9.82
	3	.25	38	42	9.81
		.5	39	53	9.81
		.65	39	53	9.82
		.8	38	39	9.82
	4	.25	40	43	9.81
		.3	60	65	9.81
		.5	37	44	9.82
		.8	25	50	9.82
	5	.2	54	50	9.81
		.25	63	63	9.81
		.5	33	42	9.81
		.8	17	57	9.81
	6	.15	50	53	9.81
		.25	59	48	9.82
		.5	25	42	9.82
		.8	11	66	9.82
$W = 10$ tons	2	.4	55	64	9.81
		.5	56	76	9.81
		.65	53	100	9.82
		.8	58	108	9.82
	3	.25	83	94	9.81
		.5	86	110	9.81
		.65	92	67	9.82
		.8	67	88	9.82
	4	.25	87	93	9.81
		.5	77	77	9.82
		.65	53	91	9.82
		.8	42	97	9.82
	5	.2	80	84	9.81
		.25	87	95	9.81
		.5	55	81	9.82
		.8	22	119	9.82
	6	.15	72	79	9.81
		.2	86	94	9.81
		.5	37	83	9.82
		.8	13	134	9.82

Table 9.3. (*Continued*)

	$d/W^{1/3}$	z/d	A(ft)	R(ft)	Corr. Equ.
$W = 100$ tons	2	.4	138	138	9.81
		.5	131	145	9.81
		.65	126	205	9.82
		.8	123	233	9.82
	3	.3	202	228	9.81
		.4	180	209	9.81
		.65	143	156	9.82
		.8	106	151	9.82
	4	.2	210	238	9.81
		.4	179	189	9.81
		.6	93	153	9.82
		.8	51	196	9.82
	5	.15	184	204	9.81
		.25	183	199	9.81
		.5	82	145	9.82
		.65	40	187	9.82
	6	.125	180	192	9.81
		.15	206	216	9.81
		.25	179	197	9.81
		.5	48	170	9.81
$W = 1$ KT	2	.35	283	312	9.81
		.5	288	316	9.81
		.65	297	387	9.81
		.8	239	493	9.82
	3	.25	364	409	9.81
		.4	334	345	9.81
		.6	304	386	9.81
		.8	128	319	9.82
	4	.2	380	421	9.81
		.25	352	361	9.81
		.4	293	397	9.81
		.6	112	288	9.82
	5	.15	359	391	9.81
		.2	365	355	9.81
		.25	293	404	9.81
		.4	186	255	9.82
	6	.15	382	382	9.81
		.2	334	352	9.81
		.25	277	364	9.81
		.4	102	283	9.82

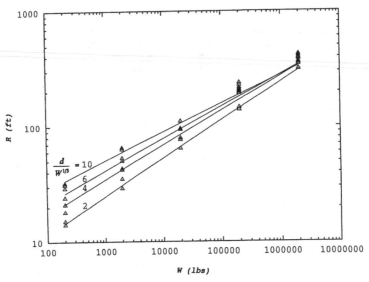

Fig. 9.13. Shallow water crater radius plotted as a function of yield.

The computed results of R are plotted in Figure 9.13 as a function of yield W. This plot indicates that the crater radius R varies as $W^{1/3}$ when $d/W^{1/3}$ is small, but as $W^{1/4}$ when $d/W^{1/3}$ becomes large. The effect of charge position on crater radius R, however, is not apparent. Plotted over the data are the approximate relations of R in terms of $d/W^{1/3}$. They are approximated by:

$$R = 1.422 \left(\frac{d}{W^{1/3}} \right)^{0.8} W^{\alpha}, \tag{9.86}$$

where

$$\alpha = 0.5 - 0.141 \left(\frac{d}{W^{1/3}} \right)^{0.25}, \quad 2 < \frac{d}{W^{1/3}} < 10. \tag{9.87}$$

This relation is approximated by assuming that the crater radius R approaches the deep water relationship (9.83) as a limit when $d/W^{1/3} \to 10$, which is the basis used in the deep water BIEM computations.

As in the deep water case, Equation (9.86) is taken to be the correlation for R, and the results in Table 9.3 are converted according to a single model given by Equation (9.81). The converted results are presented in Table 9.4.

Table 9.4. Energy equivalent crater forms adjusted in accordance with (9.81) and (9.86).

	$d/W^{1/3}$	z/d	A(ft)	R(ft)
$W = 200$ lbs	2	.35	13	14
		.5	14	
		.6	17	
		.8	16	
	3	.25	22	18
		.5	24	
		.65	22	
		.8	21	
	4	.2	24	21
		.25	43	
		.5	24	
		.8	18	
	5	.2	33	24
		.25	39	
		.5	14	
		.8	11	
	6	.15	20	26
		.2	37	
		.5	10	
		.8	7	
$W = 1$ ton	2	.4	27	31
		.5	29	
		.65	21	
		.8	24	
	3	.25	42	38
		.5	54	
		.65	44	
		.8	32	
	4	.25	40	43
		.3	91	
		.5	31	
		.8	24	
	5	.2	47	47
		.25	84	
		.5	24	
		.8	17	
	6	.15	53	50
		.25	46	
		.5	17	
		.8	12	

Table 9.4. (*Continued*)

	$d/W^{1/3}$	z/d	A(ft)	R(ft)
$W = 10$ tons	2	.4	53	67
		.5	64	
		.65	65	
		.8	76	
	3	.25	100	78
		.5	121	
		.65	65	
		.8	62	
	4	.25	94	86
		.5	56	
		.65	46	
		.8	39	
	5	.2	74	91
		.25	91	
		.5	40	
		.8	23	
	6	.15	59	96
		.2	84	
		.5	26	
		.8	15	
$W = 100$ tons	2	.4	131	145
		.5	131	
		.65	145	
		.8	161	
	3	.3	286	161
		.4	234	
		.65	113	
		.8	81	
	4	.2	292	171
		.4	198	
		.6	68	
		.8	48	
	5	.15	211	178
		.25	205	
		.5	55	
		.65	34	
	6	.125	189	183
		.15	243	
		.25	193	
		.5	36	

Table 9.4. (*Continued*)

	$d/W^{1/3}$	z/d	A(ft)	R(ft)
$W = 1$ KT	2	.35	284	311
		.5	293	
		.65	370	
		.8	309	
	3	.25	447	333
		.4	346	
		.6	352	
		.8	100	
	4	.2	466	343
		.25	370	
		.4	339	
		.6	77	
	5	.15	403	348
		.2	372	
		.25	340	
		.4	111	
	6	.15	418	349
		.2	337	
		.25	289	
		.4	68	

Table 9.4 clearly reveals two interesting points:

(1) the variation of crater height A is not sensitive to charge position z/d when $d/W^{1/3}$ is small,
(2) the crater height A diminishes rapidly as charge depth ratio z/d increases when $d/W^{1/3}$ is large.

Point 1 is the well-recognized fact that waves generated in shallow-water explosions are insensitive to the charge submergence. Point 2 indicates, however, the effect of burst depth emerges when the water depth d is not too shallow relative to the yield scale $W^{1/3}$. In other words, the effect of $z/W^{1/3}$ or the product of $d/W^{1/3}$ and z/d must be considered when $d/W^{1/3} > 2$.

For correlation purposes, data of crater height A presented in Table 9.4 are translated into a reference value B on the basis of deep water crater radius R_0 instead of R given by Equation (9.86). Values of B are presented in Table 9.5. These values can be approximated by the following relation:

$$B = \left[2.21\left(\frac{d}{W^{1/3}}\right)^{0.75} - 2.97\right]W^{\beta}, \tag{9.88}$$

where

$$\beta = 0.49\left(\frac{d}{W^{1/3}}\right)^{-0.29}, \quad 2 < \frac{d}{W^{1/3}} < 10.$$

For $d/W^{1/3} = 2$, 4, 6, and 10, the values of B are plotted in Figure 9.14. This plot shows that B varies as $W^{1/4}$ as $d/W^{1/3}$ approaches 10, the same relationship of A_0 as given by Equation (9.84). The correlation of crater height A may be obtained by

$$A = B\left(\frac{R_0}{R}\right), \tag{9.89}$$

where the depth independent radius R_0 is given by Equation (9.83) and the $d/W^{1/3}$ dependent R is given by Equation (9.86). The crater height A is approximately constant with respect to the charge position z/d for $d/W^{1/3} \leq 2$. For larger $d/W^{1/3}$, the effect of charge position z may be estimated in accordance with Equation (9.85).

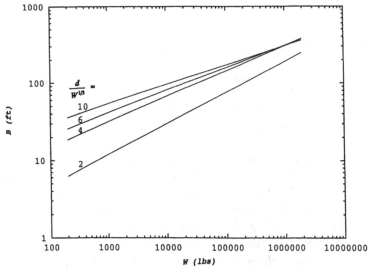

Fig. 9.14. Value B plotted as a function of yield.

Table 9.5. Energy equivalent Reference Value B.

	$d/W^{1/3}$	z/d	B(ft)	R_0(ft)
$W = 200$ lbs	2	.35	5	34
		.5	6	
		.6	7	
		.8	7	
	3	.25	12	34
		.5	13	
		.65	12	
		.8	11	
	4	.2	15	34
		.25	27	
		.5	15	
		.8	11	
	5	.2	23	34
		.25	28	
		.5	10	
		.8	8	
	6	.15	15	34
		.2	28	
		.5	8	
		.8	5	
$W = 1$ ton	2	.4	14	60
		.5	15	
		.65	11	
		.8	12	
	3	.25	27	60
		.5	34	
		.65	28	
		.8	20	
	4	.25	29	60
		.3	65	
		.5	22	
		.8	17	
	5	.2	37	60
		.25	66	
		.5	19	
		.8	13	
	6	.15	44	60
		.25	38	
		.5	14	
		.8	10	

Table 9.5. (*Continued*)

	$d/W^{1/3}$	z/d	B(ft)	R_0(ft)
$W = 10$ tons	2	.4	33	107
		.5	40	
		.65	41	
		.8	48	
	3	.25	73	107
		.5	88	
		.65	47	
		.8	45	
	4	.25	76	107
		.5	45	
		.65	37	
		.8	31	
	5	.2	63	107
		.25	77	
		.5	34	
		.8	20	
	6	.15	53	107
		.2	75	
		.5	23	
		.8	13	
$W = 100$ tons	2	.4	100	190
		.5	100	
		.65	111	
		.8	123	
	3	.3	242	190
		.4	198	
		.65	96	
		.8	69	
	4	.2	263	190
		.4	178	
		.6	61	
		.8	43	
	5	.15	198	190
		.25	192	
		.5	51	
		.65	32	
	6	.125	182	190
		.15	234	
		.25	186	
		.5	35	

Table 9.5. (*Continued*)

	$d/W^{1/3}$	z/d	B(ft)	R_0(ft)
$W = 1$ KT	2	.35	262	337
		.5	270	
		.65	341	
		.8	285	
	3	.25	442	337
		.4	342	
		.6	348	
		.8	99	
	4	.2	474	337
		.25	377	
		.4	345	
		.6	78	
	5	.15	416	337
		.2	384	
		.25	351	
		.4	115	
	6	.15	433	337
		.2	349	
		.25	299	
		.4	70	

8. Summary

This chapter presented the theoretical basis and a numerical procedure for the simulation of subsurface explosion based on BIEM, the Boundary Integral Equation Method. Numerical results generated by using the BIEM code are also presented.

The computation of the initial bubble forms is conducted parametrically to cover a systematic variation of yield W, water depth d, and depth of burst z. The range of coverage for the yield is from 200 pounds to 1 KT. Correlation relationships for the crater radius R and the crater height A are obtained as a result of the parametric analysis.

The analysis covers both deep and shallow water events, for depth-yield ratios from $d/W^{1/3} = 2$ to $d/W^{1/3} \approx 10$. The results are limited to a range of depths of burst (DOB) below a certain critical layer under the free surface. In terms of submergence-yield ratio, the correlation covers from $z/W^{1/3} = 0.75$

to $z/W^{1/3} = 2.5$. For shallow water events, the burst depth ratio z/d is a convenient parameter for analyzing the charge position effect in water of a fixed depth. The combined effect of z/d and $d/W^{1/3}$ provides the means to analyze the charge position influence in different water depths. Nevertheless, the charge position has little effect in shallow water when $d/W^{1/3}$ is less than 2.

The analysis confirms that the crater height A varies as $W^{1/4}$ as suggested by the linear wave calibration (Chapter 3). The linear wave calibration suggests the crater radius R varies as $W^{1/3}$, however. The BIEM results show that R is proportional to $W^{1/3}$ only in shallow water. The crater radius is proportional to $W^{1/4}$ in deep water. In addition, the BIEM results show that a large-yield burst behaves quite differently from a small yield.

In particular, the effect of water depth on large yield events is not as strong as that on small yield explosions. The parameter to characterize the water depth is the depth-yield ratio $d/W^{1/3}$. The effect of this parameter is more evident for small yields than for large yields on both crater radius and crater height. For yields of $W \cong 1$ KT, there is little effect due to water depth if $d/W^{1/3} > 2$. On the other hand in a very small yield event, the effect of water depth remains apparent even $d/W^{1/3} > 10$, as indicated in the linear wave correlation presented in Chapter 3.

In linear wave correlations (Chapter 3), small yield experimental data mandate a deep water criterion with $d/W^{1/3} \approx 14$. Extrapolations have been applied uniformly in the linear wave correlation for small or large yields. It is evident that the linear wave correlations are biased by the experimental results of small yield with the majority of the shots ranging from 0.5 to 100 pounds (Chapter 3). With the information available from BIEM simulations, a revision of the linear wave correlation is warranted.

REFERENCES

Abramowitz, M. and I. A. Stegun (1964). *Handbook of Mathematical Functions*, Government Printing Office, Washington, DC, USA.

Allen, R. T. (1979). "Surface Wave Prediction for Explosions in Shallow Water", DNA 001-79-C-0400, Defense Nuclear Agency, Washington, DC, USA.

Amsden, A. A. (1973). "Numerical Calculations of Surface Waves. A Modified ZUNI Code with Surface Particles and Partial Cells", Report LA-5146, Los Alamos Scientific Laboratory, USA.

Ballhaus, W. F., Jr. and M. Holt (1974). "Interaction Between the Ocean Surface and Underwater Spherical Blast Waves", *The Physics of Fluid* **17**, 6:1068–1079.

Benjamin, T. B. and J. E. Feir (1967). "The Disintegration of Wave Trains on Deep Water", *J. Fluid Mech.* **27**, 3:417–430.

Biesel, F. and F. Suquet (1951). "Laboratory Wave Machine Apparatus", *La Houille Blanche* 2:147; 4:475; 5:723.

Biot, M. A. (1956). "Theory of Propagation of Elastic Waves in Fluid Saturated Porous Solid", *J. Acoustical Society of America* **28**, 2:168–191.

Biot, M. A. (1962). "Mechanics of Deformation and Acoustic Propagation in Porous Media", *J. Appl. Phys.* **32**.

Blake, J. R. and D. C. Gibson (1981). "Growth and Collapse of a Vapour Cavity Near a Free Surface", *J. Fluid Mech.* **111**:123–140.

Blake, J. R., B. B. Taib and G. Doherty (1986). "Transient Cavities Near Boundaries", *J. Fluid Mech.* **170**:479–497.

Bottin, R. R., Jr. and D. G. Outlaw (1988a). "The Vulnerability of Coastal Military Facilities and Vessels to Explosion-Generated Water Waves; Report 1: Explosion Generated Wave Measurements, Experimental Model Investigation", Technical Report CERC-88-2, US Army Engineers Waterways Experiment Station, Vicksburg, MS, USA.

Bottin, R. R., Jr. and D. G. Outlaw (1988b). "The Vulnerability of Coastal Military Facilities and Vessels to Explosion-Generated Water Waves; Report 2: Measurement of Vessel Response, Experimental Model Investigation", Technical Report CERC 88-2, US Army Engineers Waterways Experiment Station, Vicksburg, MS, USA.

Bottin, R. R., Jr. (1990). "Impulsive Waves Generated by Falling Weights in Shallow Water: Laboratory Investigation", Technical Report CERC-90-9, US Army Engineers Waterways Experiment Station, Vicksburg, MS, USA.

Bottin, R. R., Jr. and J. E. Fowler (1990). "Surface Waves Generated by Explosions in Shallow Water: Field Investigation", Technical Report CERC-90-11, US Army Engineers Waterways Experiment Station, Vicksburg, MS, USA.

Breeding, J. E., Jr. (1972). "Refraction of Gravity Waves", Report NCSL 124-72, Naval Coastal Systems Laboratory, San Diego, CA, USA.

Bretschneider, C. L. (1954). "Field Investigation of Wave Energy Loss of Shallow Water Ocean Waves", Beach Erosion Board TM. 46, Washington, DC, USA.

Carstens, M. R., R. M. Neilson and H. D. Altinkilek (1969). "Bed Forms Generated in the Laboratory under an Oscillatory Flow; Analytical and Experimental Study", Technical Memorandum No. 28, CERC, US Army Engineers Waterways Experiment Station, Vicksburg, MS, USA.

Cauchy, A. L. (1815). *Mem. Acad. Roy. Sci.* **I**:1827. Also *Oeuvres Completes* (1822), Premier Series Vol. I, p. 38.

Chahine, G. L. (1977). "Interaction Between an Oscillating Bubble and a Free Surface", *Trans. ASME, J. Fluid Eng.* **99**:709–716.

Chahine, G. L. (1982). "Experimental and Asymptotic Study of Nonspherical Bubble Collapse", *Appl. Sci. Res.* **38**:187–198.

Chahine, G. L. (1990). "Numerical Modelling of the Dynamic Behavior of Bubbles in Nonuniform Flow Fields", *Symp. Numerical Models for Multiphase Flows*, Toronto, Canada.

Chahine, G. L., T. O. Perdue and C. B. Tucker (1988). "Interaction Between an Underwater Explosion and a Solid Submerged Structure", Tracor Hydronautics Technical Report 86029-1, Lowell, MD, USA.

Chang, P., W. K. Melville and J. W. Miles (1979). "On Evolution of Solitary Wave in a Gradually Varying Channel", *J. Fluid Mech.* **95**, 3:401–414.

Chwang, A. T. and T. Y. Wu (1976). "Cylindrical Solitary Waves in Waves on Water Variable Depth", *Lecture Notes in Physics*, Vol. 64, eds. D. G. Provis and R. Radok, Springer-Verlag, Berlin, New York, pp. 80–90.

Cole, R. H. (1948). *Underwater Explosion*, Princeton University Press, New Jersey.

Collins, J. L. (1972). "Prediction of Shallow Water Spectra", *J. Geophys. Res.* **77**:2693–2707.

Craig, B. G. (1974). "Experimental Observations of Underwater Explosion Near the Surface", Report LA-5548-MS, Los Alamos National Laboratory, USA.

Divoky, D., B. Le Méhauté and A. Liu (1970). "Breaking Waves on Gentle Slopes", *J. Geophys. Res.* **75**, 9:1681–1692.

Dommermouth, D. C. and D. K. P. Yue, (1987). "Numerical Simulations of Non-Linear Axisymmetric Phase with a Free Surface", *J. Fluid Mech.* **178**:195–219.

DNA, (1978). *Effects Manual — 1. Capability of Nuclear Weapons*, Defense Nuclear Agency, Washington, DC, USA.

Eckart, C. (1948). "The Approximate Solution of One-Dimensional Wave-Equations", *Review of Modern Physics* **20**, 1:399–417.

Favre, H. (1935). "Etude Theorique et Experimentale Des Ondes De Translation Dans les Canaux Decouverts", Paris, DUNOD, 192 Rue Bonaparte (VI).

Fennema, R., and H. Chowdhry (1987). "Simulation of One-Dimensional Dam-Break Flows", *J. Hydraulic Res.* **25**, 1.

Fogel, M. B., R. T. Allen, M. L. Gittings and R. L. Bjark (1983). "Water Waves from Underwater Explosion", DNA 001-82-C-0132, Defense Nuclear Agency, Washington, DC, USA.

Gade H. G. (1958). "Effects of a Non-Rigid Impermeable Bottom on Plane Surface Waves in Shallow Water", *J. Marine Res.* **16**:61–62.

Goda, Y. (1970). "A Synthesis of Breaker Indices", *Trans. Japanese Society of Civil Engineering* **2**, 2. (Also *Shore Protection Manual* (1977). CERC, US Army Engineers Waterways Experiment Station).

Graber, H. C. and O. S. Madsen (1988). "A Finite Depth Wind Wave Model", *J. Phys. Ocean* **18**, 11:1465–1483.

Grant, W. D. (1977). "Bottom Friction Under Waves in the Presence of Current," Ph.D. Thesis, MIT, USA.

Grant, W. D. and O. S. Madsen (1982). "Moveable Bed Roughness in Unsteady Oscillatory Flow", *J. Geophys. Res.* **87**, C1:469–481.

Grant, W. D.and O. S. Madsen (1986). "The Continental Shelf Boundary Layer", *Ann. Rev. Fluid Mechanics* **18**:265–305.

Hasselman K. and J. L. Collins (1968). "Spectral Dissipation of Finite Depth Gravity Waves Due to Turbulent Bottom Friction", *J. Marine Res.* **26**, 1:1–12.

Hirt, C. W. and W. C. Rivard (1983). "Wave Generation from Explosively Formed Cavities", DNA-001-82-C-0108, Defense Nuclear Agency, Washington, DC, USA.

Holt (1977). "Underwater Explosions", *Ann. Rev. Fluid Mechanics* **9**: 187–214.

Houston, J. R. and L. W. Chou (1983). "Numerical Modeling of Explosion Waves", Technical Report HL-83-1, CERC, US Army Engineers Waterways Experiment Station, Vicksburg, MS, USA.

Iwagaki, Y. (1968). "Hyperbolic Waves and Their Shoaling", *Proc. 11th Conf. on Coastal Engineering*, London, UK, pp. 124–144.

Iwagaki, Y. and T. Kakimura (1967). "On the Bottom Friction of Five Japanese Coasts", *Coastal Engineering in Japan* **10**:45–53.

Iwagaki, Y and T. Sagai (1971). "Shoaling of Finite Amplitude Long Waves on a Beach of Constant Slope", *Proc. 8th Conf. Coastal Engineering*, ASCE, Mexico City, pp. 60–76.

Jeffreys, H. and B. S. Jeffreys (1956). *Methods of Mathematical Physics*, Cambridge Press: Cambridge.

Johnson, J. W. (1959). "Scale Effects in Hydraulic Models Involving Wave Motion", *Trans. AGU* **30**, 4:517–525.

Johnson, J. W. and K. J. Bermel (1949). "Impulsive Waves in Shallow Water as Generated by Falling Weights", *Trans. AGU* **30**, 2:223–230.

Johnsson, I. G. (1966). "Wave Boundary Layer and Friction Factors", *Proc. 10th Conf. Coastal Engineering*, ASCE, Vol. 1, pp. 127–148.

Jordaan, J. M., Jr. (1964). "Run-up by Impulsively Generated Water Waves", Y-F008-08-02-129. DASA-14.083, Type C, Report No. 1, DASA, Washington, DC, USA.

Kajiura, K. (1963). "The Leading Wave of a Tsunami", *Bull. Earthquake Res. Inst.* **41**:535–571.

Kajiura K. (1964). "On The Bottom Friction in an Oscillatory Current", *Bull. Earthquake Res. Inst.* **42**:147–174.

Kajiura, K. (1968). "A Model Boundary Layer in Water Waves", *Bull. Earthquake Res. Inst.* **46**:75.

Kaplan, K. and O. H. Cramer (1963). "A Study of Explosion-Generated Surface Water Waves", Report URS-162-8, United Research Company, San Mateo, CA, USA.

Katopodes, N. and T. Strelkoff (1978). "Computing Two-Dimensional Dam-Break Flood Waves", *J. Hyd. Div. Amer. Soc. Civil Engrs.* pp. 1269–1287.

Kedrinskii, V. K. (1978). "Surface Effect from an Underwater Explosion (Review)", translated from *Shurnal Prikladni Mekhankiki Technicheski Fiziki*, Vol. 4.

Khangaonkar, T. (1990). "A Dissipative Nonlinear Theory for Generation Propagation of Explosion Generated Waves in Shallow Water", Ph.D. Dissertation, University of Miami, USA.

Khangaonkar, T. and B. Le Méhauté (1991). "Original Disturbance from Wave Records", *Appl. Oc. Res.* **13**, 3:110–115.

Kishi T. (1975). ONR Report N00014-69-C-0107, Office of Naval Research, Washington, DC, USA (in Japanese).

Ko, K. and H. H. Kuehl (1979). "Cylindrical and Spherical KdV Solitary Waves", *Physics of Fluids* **22**, 7.

Kranzer, H. C. and J. B. Keller (1959). "Water Waves Produced by Explosions", *J. Appl. Physics* **30** 3:398–407.

Kreibel, A. R. (1969). "Cavities and Waves from Explosions in Shallow Water", Report URS 679-5, URS Research Company, San Mateo, CA, USA.

Lamb, H. (1904). "On Deep-Water Waves", *Proc. London Mathematics Society* 371–400.

Lamb, H. (1922). "On Water Waves Due to a Disturbance Beneath the Surface", *Proc. London Mathematics Society* **21**, 2:359–372.

Lamb, H. (1932). *Hydrodynamics*, 6th ed., Cambridge University Press, Cambridge.

Lauterborn, W. and H. Bolle (1975). "Experimental Investigations of Cavitation Bubble Collapse in the Neighbourhood of a Solid Boundary", *J. Fluid Mech.* **72**:391–399.

Le Méhauté, B. (1963). "The Principle of Superposition Applied to the Theory of Cauchy-Poisson", DA-49-146-XZ-151, DASA, Washington, DC, USA.

Le Méhauté, B. (1970). "Explosion-Generated Water Waves", *8th Symp. Naval Hydrodynamics*, Pasadena, CA, USA, pp. 71–91.

Le Méhauté, B. (1971). "Explosion-Generated Water Waves", *Advances in Hydrosciences*, Academic Press, New York.

Le Méhauté, B. (1976). *An Introduction to Hydrodynamics and Water Waves*, Springer-Verlag, Berlin, New York.

Le Méhauté, B. (1986). "Explosion-Generated Water Waves and Submarine Response", Vol. I, DNA-001-84-C-0053, Defense Nuclear Agency, Washington, DC, USA.

Le Méhauté, B. (1991). "Similitude", *The Sea, Ocean Engineering Science*, John Wiley and Sons, Inc., New York, Chpt. 29, pp. 955–980.

Le Méhauté, B. and J. Wang (1980). "Transformation of Monochromatic Waves for Deep to Shallow Water", Technical Report No. 80-2, CERC, US Army Corps of Engineering, Ft. Belvoir, VA, USA.

Le Méhauté, B., C. C. Lu and E. Ulmer (1984). "A Parameterized Solution to Nonlinear Wave Problem", *J. Wtrwys, Port, Coast. and Oc. Eng.*, ASCE, August.

Le Méhauté, B. and M. Soldate (1986). "Methodology and Formulation of the EGWW Environment", DNA TR-85-99-V8-AP-7, Defense Nuclear Agency, Washington, DC, USA.

Le Méhauté, B., S. Wang and C. C. Lu (1987). "Cavities, Domes and Spikes", *J. Hydraulic Res.* **25**, 5:583–601.

Le Méhauté, B., S. Wang, T. Khangaonkar and D. Outlaw (1989). "Advances in Impulsively Generated Water Waves", *NATO Advanced Research Workshop on Water Wave Kinematics*, 22–25 May, Melde-Norway, Kluwer Academic Publishers, London.

Le Méhauté, B. and T. Khangaonkar (1991). "The Sizing of the Water Caused by an Explosion in Shallow Water: Part I — Theory; Part II — Application", *62th Shock and Vibration Symp.*, 29–31 October, Springfield, VA, USA.

Le Méhauté, B. and T. Khangaonkar (1992). "Generation and Propagation of Explosion-Generated Waves in Shallow Water," DNA-TR-92-40, Defense Nuclear Agency, Washington, DC, USA.

Lin, W. M. (1984). "Nonlinear Motion of the Free Surface Near a Moving Body", Ph.D. Thesis, Department of Ocean Engineering, MIT, USA.

Liu, P. L. F. (1973). "Damping of Water Waves Over Porous Bed", *J. Hydraulic Division, ASCE* **99**:2263–2271.

Liu, P. L. F. and R. A. Dalrymple (1984). "The Damping of Gravity Waves due to Percolation", *Coast. Eng.* **9**.

Longuet-Higgins, M. S. (1972). "A Class of Exact Time-Dependent Free Surface Flow", *J. Fluid Mech.* **55**:529–543.

Longuet-Higgins, M. S. (1983). "Bubbles, Breaking Waves and Hyperbolic Jets at a Free Surface", *J. Fluid Mech.* **127**:103–121.

Longuet-Higgins, M. S. and E. D. Cokelet (1976). "The Deformation of Steep Surface Waves on Water, I. A Numerical Method of Computation", *Proc. R. Soc. London* **A350**:1–26.

Mader, C. L. (1972). "Oscillations Near the Water Surface", Report LA-4958, Los Alamos Scientific Laboratory, USA.

Mader, C. L. (1988). *Numerical Modelling of Water Waves*, University of California Press, Berkeley.

Madsen, O. S. and W. D. Grant (1976). "Sediment Transport in the Coastal Environment", Technical Report 209, R. M. Parson Laboratory, MIT, Cambridge, MA, USA.

McCowan, J. (1894). "On the Highest Wave of Permanent Type", *Philosophical Magazine* **38**:351.

McIver, P. and D. H. Peregrine (1981). "Comparison of Numerical and Analytical Results for Waves that are Starting to Break", *Int. Symp. Hydrodynamics in Ocean Engineering*, Norwegian Institution of Technology, pp. 203–215.

Mei, C. C. (1983). *The Applied Dynamics of Ocean Surface Waves*, Wiley Interscience Publications (1989 edition published by World Scientific Publishing, Singapore).

Miche, R. (1944). "Movement Ondulatoires de la Mer en Profondeur Constante ou Decroissante", *Ann. Des Ponts et Chaussees*, pp. 25–75, 131–164, 270–292, 369–406.

Milne, W. E. (1953). *Numerical Solution of Differential Equations*, John Wiley and Sons, New York.

Miles, J. W. (1978). "An Axisymmetric Boussinesq Wave", *J. Fluid Mech.* **84**.

Moulton, J. F., Jr. and B. Le Méhauté (1980). "EGWW", DASIAC-TN-81-5, Defense Nuclear Agency, Washington, DC, USA.

Munk, W. H. and R. S. Arthur (1951). "Wave Intensity Along a Refracted Ray of Gravity Waves", *National Bureau of Standards Circular* 521:95–108.

Newman, J. N. (1985). "Transient Axisymmetric Motion of a Floating Cylinder", *J. Fluid Mech.* **157**:17–33.

Outlaw, D. G. (1984). "A Portable Directional Irregular Wave Generator for Wave Basins", *Symp. Description and Modelling of Directional Seas*, 18–20 June, Technical University of Denmark, Paper No. B-3.

Pace, C. E., R. W. Whalin and J. N. Strange (1970). "Surface Waves Resulting from Explosions in Deep Water", Technical Report No. 647, Report 5, US Army Engineers Waterways Experiment Station, Vicksburg, MS, USA.

Peregrine, D. H. (1967). "Long Waves on Beach", *J. Fluid Mech.* **27**: 815–827.

Pierson, W. J., J. J. Tuttell and J. A. Woodley (1953). "The Theory of Refraction of a Short Crested Gaussian Sea Surface with Application to the Northern New Jersey Coast", *Proc. 3rd Conf. Coastal Engineering*, pp. 86–108.

Pinkston, J. M., Jr. (1964). "Surface Waves Resulting from Explosions in Deep Water; Summary of Experimental Procedures and Results of Tests at WES Underwater Explosion Test Site", Technical Report No. 1–647, Report 1, US Army Engineers Waterways Experiment Station, Vicksburg, MS, USA.

Pinkston, J. M., Jr. and F. W. Skinner (1970). "Mono Lake Explosion Test Series, 1965, Results of Surface Wave Experiments", Technical Report N-70-5, February, US Army Engineers Waterways Experiment Station, Vicksburg. MS, USA.

Plesset, M. S. and R. B. Chapman (1971). "Collapse of an Initially Spherical Vapour Cavity in the Neighborhood of a Solid Boundary", *J. Fluid Mech.* **47**:283–290.

Poisson, S. D. (1816). *Mem. Acad. Roy. Sci.* **I**.

Prosperetii, A. (1982). "Bubble Dynamics: A Review and Some Recent Results", *Appl. Sci. Res.* **38**:145–164.

Putnam, J. A. (1949). "Loss of Wave Energy Due to Percolation in a Permeable Sea Bottom", *Trans. American Geophysical Union* **38**, 3.

Rayleigh, L. (1917). "On the Pressure Developed During the Collapse of a Spherical Cavity", *Phil. Mag.* **34**:94–97.

Reid, R. O. and K. Kajiura (1957). "On the Damping of Gravity Waves over a Permeable Bed", *Trans. American Geophysical Union* **38**, 5.

Sakkas, J. G. and T. Strelkoff (1973). "Dam-Break Flood in a Prismatic Dry Channel", *J. Hyd. Div., ASCE* **99**, HY12:2195–2216.

Schmidt, R. M. and K. A. Holsapple (1980). "Theory and Experiments on Centrifuge Cratering", *J. Geophys. Res.* **85**, B1:235–252.

Schmidt, R. M., K. A. Holsapple and K. R. Housen (1986). "Gravity Effects in Cratering", DNA 001-82-C-0301, Defense Nuclear Agency, Washington, DC, USA.

Schmidt, R. M. and K. R. Housen (1987). "Centrifuge Water Burst Experiments", Technical Report DNA-TR-87-75, Defense Nuclear Agency, Washington, DC, USA.

Shemdin, D. H., J. E. Blue and J. A. Dume (1975). "SEASAT Surface Truth Program, Maryland Test", Jet Project Laboratory, Pasadena, CA, USA, pp. 622–625.

Shima, A. and Nakajima K. (1977). "The Collapse of a Non-Hemispherical Bubble Attached to a Solid Wall", *J. Fluid Mech.* **80**:369–391.

Shore Protection Manual (1977). CERC, US Army Engineers Waterways Experiment Station, USA.

Snay, H. G. (1966). "Hydrodynamic Concepts, Selected Topics for Underwater Nuclear Explosions", TR 65-52, Naval Ordnance Laboratory, Washington, DC, USA.

Sneddon, I. N. (1951). *Fourier Transforms*, McGraw-Hill Book Co., New York.

Soldate M. and B. Le Méhauté (1986). "Propagation of Transient Wave on Nonuniform Bathymetrey", EGWW and Submarine Response, Vol. 4, DNA 001-84-C-0053, Defense Nuclear Agency, Washington, DC, USA.

Stoker, J. H. (1957). *Water Waves*, Interscience Publishers Inc., New York.

Svendsen, I. A. and O. Brink-Kjaer (1972). "Shoaling of Conoidal Waves", *Proc. 13th Coastal Engineering Conf.*, ASCE, pp. 365–383.

Swan, C. (1984). "The Damping of Surface Gravity Waves over Natural Sea Bed", M.Sc. Thesis, University of Miami, USA.

Takahashi, S. (1982). "Simple Method for Water Wave Damping by Sea-Seabed Interactions", M.Sc. Thesis, University of Miami, USA.

Tappert, F. and C. N. Judice (1972). "Recurrence of Nonlinear Ion Acoustic Waves", *Physical Review Letters* **29**:19.

Terazawa, K. (1915). "On Deep Sea Water Waves Caused by a Local Disturbance on or Beneath the Surface", *Proc. Roy. Soc. London* **92**, A: 57–81.

Trowbridge J. and O. S. Madsen (1984). "Turbulent Wave Boundary Layers", *J. Geophys. Res.* **89**, C5:7989–7997.

Unoki, S. and M. Nakano (1953a). "On the Cauchy-Poisson Waves Caused by the Eruption of a Submarine Volcano, Paper I", *Oceanographical Magazine* **4**, 4:119–141.

Unoki, S. and M. Nakano (1953b). "On the Cauchy-Poisson Waves Caused by the Eruption of a Submarine Volcano, Paper II", *Oceanographical Magazine* **5**, 1:1–13.

Unoki, S. and M. Nakano (1953c). "On the Cauchy-Poisson Waves Caused by the Eruption of a Submarine Volcano, Paper III", *Met. and Geophys.* **4**, 3 & 4:139–150.

Van Dorn, W. G. (1964). "Explosion-Generated Waves in Water of Variable Depth", *J. Mar. Res.* **22**, 2:123–141.

Van Laperen, M. P. (1975). "The Bottom Friction of the Sea-bed Off Melkbosstrand", *South Africa Dtsch Hydrogr. Zr.* **28**:72–88.

Van Mater, P. R. and E. Neal (1970). "On the Prediction of Impulsively Generated Waves Propagating into Shallow Waters", *8th Symp. Nav. Hydrodynamics*, Pasadena, CA, USA.

Vitale, P. (1979). "Sand Bed Friction Factor for Oscillatory Flows", *J. Wtrwys, Port, Coast. and Oc. Eng. ASCE* **105**, WW3.

Wang, S. (1987). "The Propagation of the Leading Wave", *ASCE Specialty Conf. Coastal Hydrodynamics*, University of Delaware, USA, pp. 657–670.

Wang, S. (1992). "Simulation of Explosion Bubble by BIEM", *63rd Shock and Vibration Symp.*, Las Cruces, New Mexico, USA.

Wang, S., R. Wade and R. Wier (1977). "Vulnerability of Surface Effect Vehicles to EGWW", ONR, N00014-76-C-0261, Office of Naval Research, Washington, DC, USA.

Wang, S. and B. Le Méhauté (1987). "Water Waves Generated by Subsurface Explosions in Very Shallow Water", DNA-TR-No. 001-85-C-0175, Defense Nuclear Agency, Washington, DC, USA.

Wang, S., B. Le Méhauté and C. C. Lu (1988). "Effect of Dispersion on Impulsive Waves", *J. Marine Geophys. Res.* **9**, 1:95–111.

Wang, S., B. Le Méhauté, T. Khangaonkar and D. Outlaw (1989). "Inverse Correlation of Explosion Generated Water Waves", *60th Shock and Vibration Symp.*, Virginia Beach, VA, USA, Vol. 5:117–147.

Wang, S. and S. Nadiga (1990). "Dynamics of Bubble Expansion and Crater Formation", DNA-TR-91-226, Defense Nuclear Agency, Washington, DC, USA.

Wang, S., B. Le Méhauté and T. Khangaonkar (1991). "Explosion-Generated Water Waves: Theory Generalization and Calibration", DNA-TR-90-108, Defense Nuclear Agency, Washington, DC, USA.

Waterways Experiment Station (1955). "Effect of Explosion in Shallow Water", US Army Engineers Waterways Experiment Station, Vicksburg, MS, USA.

Waterways Experiment Station (1986). "Shallow Water Explosion Experiments, Field Investigation by Robert Bottin, Jr. and Douglas Outlaw", Technical Report CERC-86, US Army Engineers Waterways Experiment Station, Visksburg, MS, USA.

Waterways Experiment Station (1988). "Intermediate Water Depth Explosion Experiments, Field Investigation by Robert Bottin, Jr. and Douglas Outlaw", Technical Report CERC-87-13, US Army Engineers Waterways Experiment Station, Vicksburg, MS, USA.

Waterways Experiment Station (1989). "Intermediate Water Depth Explosion Experiments, Field Investigation by Robert Bottin, Jr. and Jimmy Fowler", Technical Report CERC-89, US Army Engineers Waterways Experiment Station, Vicksburg, MS, USA.

Weidman, P. D. and R. Zakhem (1988). "Cylindrical Solitary Waves", *J. Fluid Mech.* **191**:557–573.

Whalin, R. W. (1965). "Water Waves Produced by Underwater Explosion. Propagation Theory for Regions Near the Explosion", *J. Geophys. Res.* **70**, 22:5541–5549.

Whalin, R. W. (1966). "Water Waves Generated by Shallow Water Explosions", National Science Co. Report S-359, Washington, DC, USA.

Whalin, R. W., C. E. Pace and W. F. Lane (1970). "Mono Lake Explosion Test Series, 1965. Analysis of Surface Wave and Wave Run-Up Data", Technical Report N-70-12, US Army Engineers Waterways Experiment Station, Vicksburg, MS, USA.

Whiteham, G. B. (1974). *Linear and Nonlinear Waves*, Wiley Interscience.

Wilkerson, S. (1988). "Boundary Integral Technique for Explosion Bubble Collapse Analysis", *59th Symp. Shock and Vibrations*, Albuquerque, New Mexico, pp. 209–236.

Willey, R. L. and Phillips, D. E. (1968). "Surface Phenomena Measurements and Experimental Procedures in 4000 lb HBX-1 Shallow Underwater Explosion Tests (Project Heat)", Technical Report NOLTR68-74, Naval Ordnance Laboratory, Washington, DC, USA.

Yamamoto, T., J. L. Koning, H. Sellmeijer and E. van Hijum (1978). "On the Response of a Poro-Elastic Bed to Water Waves", *J. Fluid Mech.* **87**, 1.

Yamamoto, T. (1983). "On the Response of a Coulomb-Damped Poro-Elastic Bed to Water Waves", *Marine Geotechnology* **5**, 2.

SUBJECT INDEX

AUTHOR INDEX

World Scientific
www.worldscientific.com

2587 sc

ISBN-13 978-981-02-2132-4(pbk)
ISBN-10 981-02-2132-0(pbk)

9 789810 221324